SURVIVAL ANALYSIS

RUPERT G. MILLER, JR.

Notes by
GAIL GONG

Problem Solutions by
ALVARO MUÑOZ

Stanford University

JOHN WILEY & SONS
New York   Chichester   Brisbane   Toronto   Singapore

Library of Congress Cataloging in Publication Data:

Main entry under title:
Miller, Rupert G.
   Money-making advertising.

   (Wiley series in probability and mathematical
statistics)
   Includes bibliographical references and index.
   1.   Failure time data analysis.   I. Gong, Gail.
II. Muñoz, Alvaro.   III. Title.   IV. Series.
QA273.M552   519.2   81-4437
ISBN 0-471-09434-X   AACR2

Printed in the United States of America

10 9 8 7 6 5 4 3 2 1

PREFACE

In the spring of 1980 the final quarter of a graduate
course on applied statistics at Stanford University
was devoted to the study of techniques used in analyzing survival data. Brad Efron suggested that it
would be worthwhile writing notes for these lectures,
and the contents herein represent that effort.

A series of Northern California Oncology Group se-
minars on survival analysis and a course outline from
Art Peterson at the University of Washington were
helpful in developing the syllabus for this course.
Bill Brown gave assistance and encouragement along
the way. Jerry Halpern and Terry Therneau contributed some valuable comments, and Elaine Ung pointed
out a number of misprints and inaccuracies in the
notes.

Karola Decleve did a superb job of quickly but
carefully typing the notes to keep up with the lectures and then expertly retyping them for publication.
Maria Jedd provided excellent drawings.

This work was partially supported by the National
Institute of General Medical Sciences Research Grant
GM21215.

Rupert G. Miller, Jr.
Gail Gong
Alvaro Munoz

Stanford, California
July 1981

CONTENTS

    1. Generalized Gehan Test (Breslow), 107

        Types of Tests, 109

        1.1. Permutation Covariance Matrix, 112
        1.2. Distribution Under H_0 , 113

    2. Generalized Mantel-Haenszel Test (Tarone
        and Ware), 114

        Types of Tests, 116

6. Nonparametric Methods: Regression   119

    1. Cox Proportional Hazards Model, 119

        1.1. Conditional Likelihood Analysis, 122
        1.2. Justification of the Conditional
              Likelihood, 127

              Marginal Likelihood for Ranks, 127
              Partial Likelihood, 130

        1.3. Justification of Asymptotic
              Normality, 132
        1.4. Estimation of $S(t; \underset{\sim}{x})$, 133
        1.5. Discrete or Grouped Data, 136
        1.6. Time Dependent Covariates, 139
        1.7. Example 1. Stanford Heart
              Transplant Data, 140
        1.8. Example 2. Adoption and
              Pregnancy, 141

    2. Linear Models, 141

        Accelerated Time Models, 142

        2.1. Linear Rank Tests, 143
        2.2. Least Squares Estimators, 146

              Miller Estimators, 146
              Buckley-James Estimator, 150
              Koul-Susarla-Van Ryzin
              Estimator, 154

Survival Analysis

ONE

INTRODUCTION TO SURVIVAL CONCEPTS

Survival analysis is a loosely defined statistical
term that encompasses a variety of statistical techniques for analyzing positive-valued random variables. Typically the value of the random variable is
the time to the failure of a physical component (me-
chanical or electrical) or the time to the death of a
biological unit (patient, animal, cell, etc.). However, it could be the time to the learning of a
skill, or it may not even be a time at all. For example, it could be the number of dollars that a
health insurance company pays in a particular case.
In some cases, the patient's illness is over, and the
total claim is known. In other cases, the patient is
still sick, and only the amount of the claim paid to
date is known.

While the origins of survival analysis might be
attributed to the early work on mortality tables
centuries ago, the more modern era started about a
half century ago with applications to engineering.
World War II stimulated interest in the reliability
of military equipment, and this interest in reliability carried over into the postwar era for military

and commercial products. Most of the statistical research for engineering applications was concentrated
on parametric models. Within the past two decades
there has been an increase in the number of clinical
trials in medical research, and this has shifted the
statistical focus to nonparametric approaches. While
these lecture notes attempt to cover both the parametric and nonparametric methods, the emphasis is on
the more recent nonparametric developments with applications to medical research.

REFERENCE
Leavitt and Olshen, unpublished report (1974),
    give the insurance example.

1. SURVIVAL FUNCTIONS AND HAZARD RATES

Let $T \geq 0$ have density $f(t)$ and distribution
function (d.f.) $F(t)$. The survival function $S(t)$
is

$$S(t) = 1 - F(t) = P\{T > t\} ,$$

and the hazard rate or hazard function $\lambda(t)$ is

$$\lambda(t) = \frac{f(t)}{1 - F(t)} .$$

(In epidemiology $\lambda(t)$ was historically called the
force of mortality.) The hazard rate has the interpretation

$$\lambda(t)dt \cong P\{t < T < t + dt | T > t\}$$

$$= P\begin{Bmatrix} \text{expiring in the} & \text{survived} \\ \text{interval } (t, \ t+dt) & \text{past time } t \end{Bmatrix} .$$

Integrating $\lambda(t)$,

$$\int_0^t \lambda(u)\,du = \int_0^t \frac{f(u)}{1 - F(u)}\,du = -\log[1 - F(u)]\Big|_0^t ,$$

$$= -\log[1 - F(t)] = -\log S(t) ,$$

which leads to the important expression

$$S(t) = e^{-\int_0^t \lambda(u)\,du} .$$

Notice that $F(+\infty) = 1$ (i.e., $S(+\infty) = 0$) iff $\int_0^\infty \lambda(u)\,du = +\infty$.

Note that the above concepts can be extended to the case when T does not have a density, that is, when the distribution function F has jumps. Our convention is to assume continuity, but to modify concepts and formulas to include jumps in the distribution function when it is important to do so.

2. TYPES OF CENSORING

Everything we have talked about so far can be found in any basic statistics course. What distinguishes survival analysis from other fields of statistics is censoring. Vaguely speaking, a censored observation contains only partial information about the random variable of interest. We consider three types of censoring.

Let $T_1, T_2, \ldots, T_n$ be independently, identically distributed (iid) each with d.f. F.

2.1. Type I Censoring

Let t_c be some (preassigned) fixed number which we call the fixed censoring time. Instead of observing $T_1, \ldots, T_n$ (the random variables of interest) we can only observe $Y_1, \ldots, Y_n$ where

$$Y_i = \begin{cases} T_i & \text{if } T_i \leq t_c \text{ ,} \\ t_c & \text{if } t_c < T_i \text{ .} \end{cases}$$

Notice that the distribution function of Y has positive mass $P\{T > t_c\} > 0$ at $y = t_c$.

2.2. Type II Censoring

Let $r < n$ be fixed, and let $T_{(1)} < T_{(2)} < \cdots < T_{(n)}$ be the order statistics of $T_1, T_2, \ldots, T_n$. Observation ceases after the r-th failure so we can observe $T_{(1)}, \ldots, T_{(r)}$. The full ordered observed sample is

$$Y_{(1)} = T_{(1)}$$
$$\vdots$$
$$Y_{(r)} = T_{(r)}$$
$$Y_{(r+1)} = T_{(r)}$$
$$\vdots$$
$$Y_{(n)} = T_{(r)} \text{ .}$$

Both Type I and Type II censoring arise in engineering applications. In such situations there is a batch of transistors or tubes; we put them all on test at $t = 0$, and record their times to failure. Some transistors may take a long time to burn out, and we will not want to wait that long to end the experiment. Therefore, we might stop the experiment at a prespecified time t_c, in which case we have Type I censoring, or we might not know beforehand

what value of the fixed censoring time is good so we
decide to wait until a prespecified fraction r/n of
the transistors has burned out, in which case we have
Type II censoring.

2.3. Random Censoring

Let C_1, C_2, ..., C_n be iid each with d.f. G.
C_i is the censoring time associated with T_i. We
can only observe (Y_1, δ_1), ..., (Y_n, δ_n) where

$$Y_i = \min(T_i, C_i) = T_i \wedge C_i ,$$

$$\delta_i = I(T_i \leq C_i) = \begin{cases} 1 & \text{if } T_i \leq C_i \text{ , that is,} \\ & T_i \text{ \underline{is not} censored,} \\ 0 & \text{if } T_i > C_i \text{ , that is,} \\ & T_i \text{ \underline{is} censored .} \end{cases}$$

Notice that Y_1, ..., Y_n are iid with some d.f. H.
Also δ_1, ..., δ_n contain the censoring information.
(In Type I and Type II censoring we also were able to
observe which items were censored, but since it was
easy to see which ones these were, we didn't need to
define the δ_i's explicitly.)

Random censoring arises in medical applications
with animal studies or clinical trials. In a clinical trial, patients may enter the study at different
times; then each is treated with one of several
possible therapies. We want to observe their life-
times, but censoring occurs in one of the following
forms:

1. Loss to follow-up. The patient may decide to
   move elsewhere; we never see him again.

2. Drop out. The therapy may have such bad side
   effects that it is necessary to discontinue
   the treatment. Or the patient may still be in
   contact (he hasn't moved), but he refuses to
   continue the treatment.

3. Termination of the study.

The following picture illustrates a possible
trial:

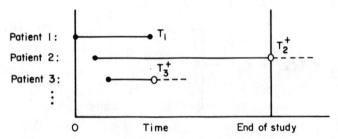

Here, patient 1 entered the study at t = 0 and died
at T_1 to give an uncensored observation; patient 2
entered the study, and by the end of the study he was
still alive resulting in a censored observation T_2^+ ;
and patient 3 entered the study and was lost to
follow-up before the end of the study to give another
censored observation T_3^+.

   With random censoring we will make the following
crucial assumption.

   Assumption: T_i and C_i are independent.

Without this assumption few results are available.
It seems justified with random entries to the study
and randomly occurring losses to follow-up. However,
if the reason for dropping out is related to the

course of the therapy, there may well be dependence between T_i and C_i.

2.4. Other Types of Censoring

There are other types of censoring which appear in the literature. The previous types of censoring fall under the heading of <u>right censoring</u>: if the random variable of interest is too large, we do not get to observe it completely. There is also <u>left censoring</u>. For example, in random left censoring, we can only observe $(Y_1, \varepsilon_1), \ldots, (Y_n, \varepsilon_n)$ where

$$Y_i = \max(T_i, C_i) = T_i \vee C_i ,$$

$$\varepsilon_i = I(C_i \leq T_i) .$$

<u>Example. African Children</u>. Here both right and left censoring are present. A Stanford psychiatrist wanted to know the age at which a certain group of African children learned to perform a particular task. When he arrived in the village, there were some children who already knew how to perform the task, so these children contributed left-censored observations. Some children learned the task while he was present, and their ages could be recorded. When he left, there remained some children who had not yet learned the task, thereby contributing to right-censored observations.

REFERENCES
Leiderman et al., <u>Nature</u> (1973).
Turnbull, <u>JASA</u> (1974).

Both right and left censoring are special cases of <u>interval censoring</u>, in which we may only get to see that the random variable of interest falls in an interval. If T_i is random right censored, we get to observe that T_i falls in the interval $[C_i, \infty)$, and if T_i is random left censored, we get to observe that T_i falls in the interval $(-\infty, C_i]$. There are examples of more general interval censoring.

In contrast to interval censoring there is truncation in which if the random variable of interest falls outside some interval, even its existence is unobserved. For example, suppose we want to get the distribution and expected size of a certain organelle in the cell. Because of limitations on the measuring equipment, if an organelle is below a certain size it cannot be detected.

Alternative Notation. We have adopted the notation that T_i is the survival time, C_i is the censoring time, and the observed random variables are $Y_i = T_i \wedge C_i$ and $\delta_i = I(T_i \leq C_i)$. There is other notation in the literature.

(i) $X_i \sim F$ is the survival time,

   $Y_i \sim G$ is the censoring time,

   $Z_i = X_i \wedge Y_i \sim H$ and $\delta_i = I(X_i \leq Y_i)$

   are the observed random variables.

This is an appealing notation, because it is easy to keep track of the random variables and the distribution functions. But we will be studying regression later, using X as the independent variable.

(ii) $X_i^0 \sim F^0$ is the survival time,

   $Y_i \sim G$ is the censoring time,

$$Z_i = X_i^0 \wedge Y_i \quad \text{and} \quad \delta_i = I(X_i^0 \leq Y_i)$$

are the observed random variables.

In reporting actual numbers, the convention is to write T_i for a noncensored observation and T_i^+ for a censored observation. Therefore, our data might consist of

5, 11+, 6.5, 14+

where the times 5 and 6.5 are not censored and 11 and 14 are censored.

TWO

PARAMETRIC MODELS

1. DISTRIBUTIONS

1.1. Exponential

The exponential model assumes constant risk:

$$\lambda(t) \equiv \lambda > 0 \ .$$

Therefore,

$$\int_0^t \lambda(u)\,du = \lambda t \ ,$$

$$S(t) = e^{-\int_0^t \lambda(u)\,du} = e^{-\lambda t} \ ,$$

$$f(t) = -\frac{d}{dt} S(t) = \lambda e^{-\lambda t} \ ,$$

$$E(T) = \frac{1}{\lambda} \ ,$$

and

$$\text{Var}(T) = \frac{1}{\lambda^2} .$$

1.2. Gamma

The gamma model is a generalization of the exponential model:

$$f(t) = \frac{\lambda^\alpha}{\Gamma(\alpha)} t^{\alpha-1} e^{-\lambda t} , \quad \alpha > 0 , \quad \lambda > 0 .$$

Then,

$$E(T) = \frac{\alpha}{\lambda}$$

and

$$\text{Var}(T) = \frac{\alpha}{\lambda^2} .$$

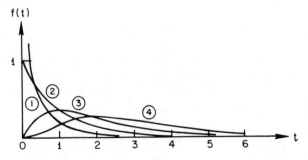

Figure 1. The gamma density for $\lambda = 1$ and
① $\alpha = 1/2$, ② $\alpha = 1$, ③ $\alpha = 2$, ④ $\alpha = 3$.

Unfortunately the gamma model does not have closed form expressions for $S(t)$ and $\lambda(t)$.

$$S(t) = 1 - \int_0^t f(u) \, du$$

$$= 1 - \left(\frac{\text{incomplete gamma function}}{\text{complete gamma function}} \right).$$

1.3. Weibull

The Weibull model is another generalization of the exponential model:

$$S(t) = e^{-(\lambda t)^\alpha} , \quad \alpha > 0 , \quad \lambda > 0 .$$

Then,

$$\int_0^t \lambda(u) \, du = (\lambda t)^\alpha ,$$

$$\lambda(t) = \alpha \, \lambda (\lambda t)^{\alpha-1} ,$$

and

$$f(t) = \lambda(t) \, S(t) = \alpha \, \lambda(\lambda t)^{\alpha-1} \, e^{-(\lambda t)^\alpha} .$$

For the Weibull model, $E(T)$ and $Var(T)$ have no nice closed form expression, but the forms of $\lambda(t)$ and $S(t)$ make the Weibull model a useful one in survival analysis. See Figure 2.

1.4. Rayleigh

$$\lambda(t) = \lambda_0 + \lambda_1 t ,$$

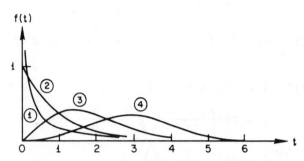

Figure 2. The Weibull density for $\textcircled{1}$ $\lambda = 1$, $\alpha = 1/2$, $\textcircled{2}$ $\lambda = 1$, $\alpha = 1$, $\textcircled{3}$ $\lambda = .5$, $\alpha = 2$, $\textcircled{4}$ $\lambda = .3$, $\alpha = 3$.

$$\int_0^t \lambda(u)\,du = \lambda_0 t + \frac{1}{2}\lambda_1 t^2 ,$$

$$S(t) = \exp(-\lambda_0 t - \frac{1}{2}\lambda_1 t^2) ,$$

and

$$f(t) = (\lambda_0 + \lambda_1 t) \exp(-\lambda_0 t - \frac{1}{2}\lambda_1 t^2) .$$

The moments have no closed form expressions. The linear risk can be generalized to polynomials:

$$\lambda(t) = \sum_{i=0}^{p} \lambda_i t^i .$$

1.5. Lognormal

Assume

$$\log T_i \sim N(\mu, \sigma^2) \ .$$

Then $\lambda(t)$ and $S(t)$ have no closed form representations.

$$S(t) = 1 - P\{T < t\} = 1 - P\{\log T < \log t\} \ ,$$

$$= 1 - P\left\{\frac{\log T - \mu}{\sigma} < \frac{\log t - \mu}{\sigma}\right\} \ ,$$

$$= 1 - \Phi\left(\frac{\log t - \mu}{\sigma}\right) \ .$$

The lognormal distribution may be convenient for use with uncensored data. A log transformation converts the data into the standard linear model setup.

1.6. Pareto

Assume

$$S(t) = \left(\frac{a}{t}\right)^\alpha I_{[a,\infty)}(t) \ , \quad \alpha > 0 \ , \quad a > 0 \ .$$

Then,

$$f(t) = \frac{\alpha \, a^\alpha}{t^{\alpha+1}} I_{[a,\infty)}(t)$$

and

$$\lambda(t) = \frac{\alpha}{t} I_{[a,\infty)}(t) \ .$$

The moments are easily calculated, but they may be infinite.

1.7.   IFR and IFRA

F or f has an <u>increasing failure rate</u> (we say
F or f is <u>IFR</u>) if $\lambda(t)$ is increasing; F or f
has an <u>increasing failure rate average</u> (we say F or
f is <u>IFRA</u>) if

$$\frac{1}{t} \int_0^t \lambda(u) \ du$$

is increasing.   Analogous definitions can be made for
DFR and DFRA.

Constant FR	IFR	DFR
Exponential	Weibull ($\alpha > 1$)	Weibull ($\alpha < 1$)
	Gamma ($\alpha > 1$)	Gamma ($\alpha < 1$)
	Rayleigh ($\lambda_1 > 0$)	Rayleigh ($\lambda_1 < 0$)
		Pareto ($t > a$)

The concepts of IFR and IFRA distributions are
useful in engineering applications, particularly in
the study of systems of components. In biostatistics
they are not usually helpful. For example, in epi-
demiological studies the risk for long-term survival
usually has a bathtub shape with time divided into
three periods as illustrated on the next page.

REFERENCE
Barlow and Proschan, <u>Statistical Theory of Re-
    liability and Life Testing</u> (1975).

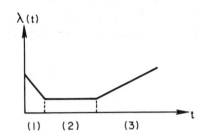

(1) Immature period

(2) Adult period

(3) Senescent period

2. ESTIMATION

2.1. Maximum Likelihood

We assume the random censoring model. (Note that this includes Type I censoring by simply setting $C_i \equiv t_c$. Also, the likelihoods for Type II censoring are similar to the ones for Type I censoring except for the multiplication of some constants to take account of the ordering.)

The pair (y_i, δ_i) has likelihood

$$L(y_i, \delta_i) = \begin{cases} f(y_i) & \text{if } \delta_i = 1 \quad \text{(uncensored)}, \\ S(y_i) & \text{if } \delta_i = 0 \quad \text{(censored)}, \end{cases}$$

$$= f(y_i)^{\delta_i} S(y_i)^{1-\delta_i},$$

and the likelihood of the full sample is

$$L = L(y_1, \ldots, y_n; \delta_1, \ldots, \delta_n) = \prod_{i=1}^{n} L(y_i, \delta_i)$$

$$= \left(\prod_u f(y_i) \right) \left(\prod_c S(y_i) \right)$$

where $\prod_u (\prod_c)$ denotes a product over the uncensored

(censored) observations. Actually, the complete
likelihoods under random censoring are

$$L(y_i, \delta_i) = \begin{cases} f(y_i)[1 - G(y_i)] & \text{if } \delta_i = 1 , \\ g(y_i) \, S(y_i) & \text{if } \delta_i = 0 , \end{cases}$$

$$L = \left(\underset{u}{\Pi} f(y_i) \right) \left(\underset{c}{\Pi} S(y_i) \right) \left(\underset{c}{\Pi} g(y_i) \right) \left(\underset{u}{\Pi} [1 - G(y_i)] \right) ,$$

but under the assumption that the censoring time has
no connection to the survival time, the last two pro-
ducts $\underset{c}{\Pi} \, g(y_i)$ and $\underset{u}{\Pi} [1 - G(y_i)]$ do not involve
the unknown lifetime parameters, so these two pro-
ducts can be treated like constants when maximizing
L.

    Let $\theta = (\theta_1, \ldots, \theta_p)'$ be the vector of para-
meters. Finding $\max_{\theta} L(\theta)$ is equivalent to finding
the solution $\hat{\theta}$ to the likelihood equations

$$0 = \frac{\partial}{\partial \theta_j} \log L(\theta) = \sum_{i=1}^{n} \frac{\partial}{\partial \theta_j} \log L_\theta(y_i, \delta_i) ,$$

$$= \underset{u}{\sum} \frac{\partial}{\partial \theta_j} \log f_\theta(y_i) + \underset{c}{\sum} \frac{\partial}{\partial \theta_j} \log S_\theta(y_i) ,$$

$$j = 1, \ldots, p .$$

Typically, calculation on a computer using iterative
methods is required.

    Newton-Raphson and Method of Scoring. Denote
$L_i(\theta) = L_\theta(y_i, \delta_i)$, $i = 1, \ldots, n$, and define

$$\frac{\partial}{\partial \theta} \log L(\theta) = \left(\frac{\partial}{\partial \theta_1} \log L(\theta), \ldots, \frac{\partial}{\partial \theta_p} \log L(\theta) \right)',$$

$$\frac{\partial^2}{\partial \underset{\sim}{\theta}^2} \log L(\underset{\sim}{\theta}) =$$

$$\begin{pmatrix} \dfrac{\partial^2}{\partial \theta_1 \partial \theta_1} \log L(\underset{\sim}{\theta}) & \cdots & \dfrac{\partial^2}{\partial \theta_1 \partial \theta_p} \log L(\underset{\sim}{\theta}) \\ \vdots & & \vdots \\ \dfrac{\partial^2}{\partial \theta_p \partial \theta_1} \log L(\underset{\sim}{\theta}) & \cdots & \dfrac{\partial^2}{\partial \theta_p \partial \theta_p} \log L(\underset{\sim}{\theta}) \end{pmatrix} .$$

Then the likelihood equations are

$$0 = \sum_i \frac{\partial}{\partial \theta_j} \log L_i(\underset{\sim}{\theta}) \ , \quad j = 1, \ \ldots, \ p \ ,$$

or

$$\underset{\sim}{0} = \frac{\partial}{\partial \underset{\sim}{\theta}} \log L(\underset{\sim}{\theta}) \ .$$

Assume $\hat{\underset{\sim}{\theta}}^0 = (\hat{\theta}_1^0, \ \ldots, \ \hat{\theta}_p^0)'$ is an initial guess at the solution. Expand about $\hat{\underset{\sim}{\theta}}^0$:

$$0 = \sum_i \frac{\partial}{\partial \theta_j} \log L_i(\hat{\underset{\sim}{\theta}}) = \sum_i \frac{\partial}{\partial \theta_j} \log L_i(\hat{\underset{\sim}{\theta}}^0)$$

$$+ \sum_k (\hat{\theta}_k - \hat{\theta}_k^0) \sum_i \frac{\partial^2}{\partial \theta_k \partial \theta_j} \log L_i(\hat{\underset{\sim}{\theta}}^0) + \ldots \ ,$$

$$j = 1, \ \ldots, \ p \ ,$$

or

$$0 = \frac{\partial}{\partial \theta} \log L(\hat{\underset{\sim}{\theta}}) = \frac{\partial}{\partial \theta} \log L(\hat{\underset{\sim}{\theta}}^0)$$

$$+ \frac{\partial^2}{\partial \theta^2} \log L(\hat{\underset{\sim}{\theta}}^0) \; (\hat{\underset{\sim}{\theta}} - \hat{\underset{\sim}{\theta}}^0) + \cdots .$$

Let $\hat{\underset{\sim}{\theta}}^1$ be the solution ignoring second order and higher terms:

$$\hat{\underset{\sim}{\theta}}^1 = \hat{\underset{\sim}{\theta}}^0 + \left(- \frac{\partial^2}{\partial \theta^2} \log L(\hat{\underset{\sim}{\theta}}^0) \right)^{-1} \frac{\partial}{\partial \theta} \log L(\hat{\underset{\sim}{\theta}}^0) . \quad (1)$$

The vector $(\partial / \partial \theta) \log L(\hat{\underset{\sim}{\theta}}^0)$ is called the <u>score</u> <u>vector</u> at $\hat{\underset{\sim}{\theta}}^0$, and the matrix

$$\underset{\sim}{i}(\hat{\underset{\sim}{\theta}}^0) = - \frac{\partial^2}{\partial \theta^2} \log L(\hat{\underset{\sim}{\theta}}^0)$$

is called the <u>sample information matrix</u> at $\hat{\underset{\sim}{\theta}}^0$. Notice that

$$E(\underset{\sim}{i}(\underset{\sim}{\theta})) = \left(-E \frac{\partial^2}{\partial \theta_k \partial \theta_j} \log L(\underset{\sim}{\theta}) \right) = \underset{\sim}{I}(\underset{\sim}{\theta}) ,$$

which is the <u>Fisher information</u>. We point out that $\underset{\sim}{I}(\underset{\sim}{\theta})$ is the Fisher information of the entire sample:

$$\underset{\sim}{I}(\underset{\sim}{\theta}) = \sum_{i=1}^{n} \underset{\sim i}{I}(\underset{\sim}{\theta}) = n \underset{\sim 1}{I}(\underset{\sim}{\theta}) ,$$

where $\underset{\sim i}{I}(\underset{\sim}{\theta})$ is the Fisher information of the i-th observation.

The iteration scheme using (1) is called the

Newton-Raphson method. Replacing the sample information in (1) by the Fisher information gives

$$\hat{\underset{\sim}{\theta}}^1 = \hat{\underset{\sim}{\theta}}^0 + \underset{\sim}{I}^{-1}(\hat{\underset{\sim}{\theta}}^0) \; \frac{\partial}{\partial\underset{\sim}{\theta}} \log L(\hat{\underset{\sim}{\theta}}^0) \; , \qquad (2)$$

and the iteration scheme using (2) is called the
method of scoring. While (2) might produce improved
convergence in some instances, it may not be possible, particularly if censoring is present, to figure
out $\underset{\sim}{I}(\underset{\sim}{\theta})$ for use in (2).

REFERENCES
Rao, Linear Statistical Inference (1965),
    Section 5g.
Gross and Clark, Survival Distributions (1975),
    Chapter 6.
Kalbfleisch and Prentice, The Statistical Analysis
    of Failure Time Data (1980), Section 3.7.

Confidence Intervals and Tests. For random and
Type I censoring, under smoothness conditions,

$$\hat{\underset{\sim}{\theta}} \overset{a}{\sim} N(\underset{\sim}{\theta}, \; \underset{\sim}{I}^{-1}(\underset{\sim}{\theta})) \; .$$

Usually for Type II censoring, this result also holds,
but the proofs are different. (The notation $\overset{a}{\sim}$
denotes "is asymptotically distributed as.")

For testing H_0: $\underset{\sim}{\theta} = \underset{\sim}{\theta}^0$ or constructing confidence intervals, we have three procedures.

(i) Wald

$$(\hat{\underset{\sim}{\theta}} - \underset{\sim}{\theta}^0)' \; \underset{\sim}{I}(\underset{\sim}{\theta}^0) \; (\hat{\underset{\sim}{\theta}} - \underset{\sim}{\theta}^0) \overset{a}{\sim} \chi^2_p \quad \text{under} \quad H_0 \; .$$

We can alternatively substitute $\underset{\sim}{I}(\hat{\underset{\sim}{\theta}})$ for
$\underset{\sim}{I}(\underset{\sim}{\theta}^0)$.

(ii) <u>Neyman-Pearson/Wilks likelihood ratio</u>

$$-2 \log \frac{L(\underset{\sim}{\theta}^0)}{L(\hat{\underset{\sim}{\theta}})} \overset{a}{\sim} \chi_p^2 \quad \text{under} \quad H_0 \;.$$

(iii) <u>Rao</u>

$$\frac{\partial}{\partial \underset{\sim}{\theta}} \log L(\underset{\sim}{\theta}^0)' \; \underset{\sim}{I}^{-1}(\underset{\sim}{\theta}^0) \; \frac{\partial}{\partial \underset{\sim}{\theta}} \log L(\underset{\sim}{\theta}^0) \overset{a}{\sim} \chi_p^2$$

$$\text{under} \quad H_0 \;.$$

Notice that Rao's method does not use the maximum likelihood estimator (MLE), so no iterative calculation is necessary. However, in addition to tests, we usually want estimates and confidence intervals, so we would need to calculate $\hat{\underset{\sim}{\theta}}$ anyway. Once we have $\hat{\underset{\sim}{\theta}}$ and $\underset{\sim}{I}(\underset{\sim}{\theta}^0)$, the Wald method is easy.

Under censoring we may need to replace $\underset{\sim}{I}(\theta)$ with $\underset{\sim}{i}(\theta)$ because calculation of $\underset{\sim}{I}(\theta)$ is usually difficult. Also, Efron and Hinkley suggest that using $\underset{\sim}{i}(\theta)$ is better than using $\underset{\sim}{I}(\theta)$ for confidence intervals even if $\underset{\sim}{I}(\theta)$ can be calculated. There is not universal agreement on this, however.

REFERENCES
Rao, <u>Linear Statistical Inference</u> (1965),
      Section 6e.
Efron and Hinkley, <u>Biometrika</u> (1978).

<u>Example 1. Exponential.</u> Under random censoring, let

$$n_u = \text{No. of uncensored observations.}$$

Then,

$$L = \lambda^{n_u} \exp\left(-\lambda \sum_u t_i - \lambda \sum_c c_i\right) ,$$

$$= \lambda^{n_u} \exp\left(-\lambda \sum_{i=1}^{n} y_i\right) ,$$

$$\log L = n_u \log \lambda - \lambda \sum_{i=1}^{n} y_i ,$$

$$\frac{\partial}{\partial \lambda} \log L = \frac{n_u}{\lambda} - \sum_{i=1}^{n} y_i ,$$

$$\hat{\lambda} = \frac{n_u}{\sum_{i=1}^{n} y_i} ,$$

$$\frac{\partial^2}{\partial \lambda^2} \log L = \frac{-n_u}{\lambda^2} ,$$

$$i(\underset{\sim}{\lambda}) = \frac{n_u}{\lambda^2} .$$

We remark that $\hat{\lambda} = n_u / \sum_{i=1}^{n} y_i$ is also the MLE under Type I and Type II censoring as well as random censoring.

To construct confidence intervals and perform tests, we need the distribution of $\hat{\lambda}$.

(a) If <u>no censoring</u> is present

$$\hat{\lambda} = \frac{n}{\sum_{i=1}^{n} T_i} = \frac{1}{\bar{T}} ,$$

where $T_1, \ldots, T_n$ are iid each with the exponential distribution

$$f_{T_1}(t) = \lambda\, e^{-\lambda t} \, .$$

Consequently, $S = \Sigma_{i=1}^{n} T_i$ has the gamma density

$$f_S(t) = \frac{\lambda^n}{\Gamma(n)} \, t^{n-1} \, e^{-\lambda t} \, ,$$

so $2\lambda \, \Sigma_{i=1}^{n} T_i \sim \chi_{2n}^2$, or equivalently,

$$\frac{2n\lambda}{\hat{\lambda}} \sim \chi_{2n}^2 \, .$$

Therefore, $2n\, \lambda/\hat{\lambda}$ is a pivotal statistic and can be used for test and confidence interval construction. (The notation "$\sim$" denotes "is distributed as.")

(b) For Type II censoring, we can rewrite

$$\sum_{i=1}^{n} Y_i = T_{(1)} + T_{(2)} + \cdots + T_{(r)} + (n-r)\, T_{(r)}$$

$$= n\, T_{(1)} + (n-1)[T_{(2)} - T_{(1)}] + \cdots$$

$$+ (n-r+1)[T_{(r)} - T_{(r-1)}] \, .$$

Using results from Poisson processes and exponential waiting times,

$$T_{(1)} = \{\text{min of } n \text{ iid exponential}(\lambda)\ T_i\text{'s}\} \, ,$$

$$\sim n\, \lambda e^{-n\lambda t} \, ,$$

$$n \, T_{(1)} \sim \lambda \, e^{-\lambda t} \, ,$$

$$T_{(2)} - T_{(1)} =$$

$$\{\min \text{ of } n-1 \text{ iid exponential}(\lambda) \, T_i\text{'s}\} \, ,$$

$$\sim (n-1)\lambda e^{-(n-1)\lambda t} \, ,$$

$$(n-1)[T_{(2)} - T_{(1)}] \sim \lambda e^{-\lambda t} \, , \text{ and so on },$$

and

$$n \, T_{(1)}, \ (n-1)[T_{(2)} - T_{(1)}], \ \ldots,$$

$$(n-r+1)[T_{(r)} - T_{(r-1)}]$$

are independent, so

$$2\lambda \sum_{i=1}^{n} Y_i \sim \chi^2_{2r} \, .$$

Thus, to construct confidence intervals and tests, $2r\lambda/\hat{\lambda}$ can be used in conjunction with a χ^2 distribution, where the degrees of freedom are twice the number of uncensored order statistics.

(c) If random or Type I censoring is present, we have no recourse but to use the asymptotic theory. As derived previously,

$$\hat{\lambda} = \frac{n_u}{\sum\limits_{i=1}^{n} y_i} \, ,$$

$$\frac{\partial^2}{\partial \lambda^2} \log L = \frac{-n_u}{\lambda^2} \, ,$$

so,

$$\frac{\hat{\lambda} - \lambda}{\sqrt{\lambda^2/n_u}} \overset{a}{\sim} N(0, 1) ,$$

where n_u may be replaced by $E(n_u)$ if the latter is available.

The normality approximation can be improved by transforming the estimate. By the delta method (to be discussed next), since

$$\hat{\lambda} \overset{a}{\sim} N\left(\lambda, \frac{\lambda^2}{n_u}\right) ,$$

then

$$\log \hat{\lambda} \overset{a}{\sim} N\left(\log \lambda, \frac{1}{n_u}\right) .$$

Notice that $1/n_u$, the asymptotic variance of $\log \hat{\lambda}$, does not depend on the unknown parameter λ . It is an empirical fact that transforming an estimate to remove the dependence of the variance on the unknown parameter tends to improve the convergence to normality by reducing the skewness.

REFERENCE
Epstein and Sobel, JASA (1953), is a classic
    paper.

Delta Method. Suppose the random variable Y has mean μ and variance σ^2 (denoted by "$Y \sim (\mu, \sigma^2)$") and suppose we want the distribution of some function $g(Y)$. Expand $g(Y)$ about μ

$$g(Y) = g(\mu) + (Y - \mu) \, g'(\mu) + \cdots$$

and ignore higher order terms to get

$$g(Y) \approx (g(\mu), \, \sigma^2(g'(\mu))^2)$$

(where "$\approx$" denotes "is approximately distributed as").
If furthermore $Y \overset{a}{\sim} N(\mu, \sigma^2)$, then

$$g(Y) \overset{a}{\sim} N(g(\mu), \, \sigma^2(g'(\mu))^2) \ .$$

The delta method also has a multivariate version.
Suppose

$$\begin{pmatrix} X \\ Y \end{pmatrix} \sim \left(\begin{pmatrix} \mu_x \\ \mu_y \end{pmatrix}, \begin{pmatrix} \sigma_x^2 & \sigma_{xy} \\ & \sigma_y^2 \end{pmatrix} \right),$$

and suppose we want the distribution of $g(X, Y)$.
Then

$$g(X, Y) = g(\mu_x, \mu_y) + (X - \mu_x) \, \frac{\partial}{\partial x} g(\mu_x, \mu_y)$$

$$+ (Y - \mu_y) \, \frac{\partial}{\partial y} g(\mu_x, \mu_y) + \cdots \ ,$$

so

$$g(X, Y) \approx \left(g(\mu_x, \mu_y), \, \sigma_x^2 (\frac{\partial}{\partial x} g)^2 \right.$$

$$\left. + 2 \, \sigma_{xy} \, \frac{\partial}{\partial x} g \, \frac{\partial}{\partial y} g + \sigma_y^2 (\frac{\partial}{\partial y} g)^2 \right) .$$

If furthermore $(X, Y) \overset{a}{\sim}$ Normal, then
$g(X, Y) \overset{a}{\sim}$ Normal.

The delta method is very useful. For example, we could use it to get an approximate value for $\mathrm{Var}(\bar{X}/\bar{Y})$ or $\mathrm{Var}(\overline{XY})$.

Example 2. Weibull. Reparametrize with $\gamma = \lambda^\alpha$ so that taking derivatives is easier:

$$S(t) = e^{-(\lambda t)^\alpha} = e^{-\gamma t^\alpha},$$

$$f(t) = \gamma\alpha\, t^{\alpha-1}\, e^{-\gamma t^\alpha}.$$

Then,

$$L = (\gamma\alpha)^{n_u} \left(\prod_u t_i^{\alpha-1} \right) \exp\left(-\gamma \sum_u t_i^\alpha \right)$$

$$\times \exp\left(-\gamma \sum_c c_i^\alpha \right),$$

$$= (\gamma\alpha)^{n_u} \left(\prod_u t_i^{\alpha-1} \right) \exp\left(-\gamma \sum_{i=1}^{n} y_i^\alpha \right),$$

$$\log L = n_u \log \gamma + n_u \log \alpha$$

$$+ (\alpha - 1) \sum_u \log t_i - \gamma \sum_{i=1}^{n} y_i^\alpha,$$

$$\frac{\partial}{\partial\gamma} \log L = \frac{n_u}{\gamma} - \sum_{i=1}^{n} y_i^\alpha,$$

$$\frac{\partial}{\partial\alpha} \log L = \frac{n_u}{\alpha} + \sum_u \log t_i - \gamma \sum_{i=1}^{n} y_i^\alpha \log y_i.$$

Therefore, the MLE $(\hat{\alpha}, \hat{\gamma})$ satisfies

$$\hat{\gamma} = \frac{n_u}{\sum\limits_{i=1}^{n} y_i^{\hat{\alpha}}} \, ,$$

$$0 = \frac{n_u}{\hat{\alpha}} + \sum\limits_{u} \log t_i - \hat{\gamma} \sum\limits_{i=1}^{n} y_i^{\hat{\alpha}} \log y_i \, .$$

These equations must be solved iteratively. The Newton-Raphson method requires the sample information matrix

$$-\frac{\partial^2}{\partial\theta^2} \log L = -\begin{pmatrix} \dfrac{\partial^2}{\partial\gamma^2} \log L & \dfrac{\partial}{\partial\gamma\partial\alpha} \log L \\[2em] & \dfrac{\partial^2}{\partial\alpha^2} \log L \end{pmatrix} \, ,$$

which is calculated in Problem 3. Also, the Newton-Raphson method requires starting values $\hat{\gamma}_0$, $\hat{\alpha}_0$. To get reasonable starting values, observe that

$$S(t) = e^{-\gamma t^{\alpha}} \, ,$$

$$\log S(t) = -\gamma t^{\alpha} \, ,$$

$$\log(-\log S(t)) = \log \gamma + \alpha \log t \, ,$$

so if we had estimates $\hat{S}(t_i)$, we could regress $\log(-\log \hat{S}(t_i))$ against $\log t_i$, and then let the regression coefficient be $\hat{\alpha}_0$ and the constant be $\log \hat{\gamma}_0$. Possible choices of $\hat{S}(t)$ are the Kaplan-Meier estimate, which we will discuss later, or the empirical distribution function, which ignores censoring.

REFERENCE
Cohen, Technometrics (1965), treats the MLE and gives additional references.

Estimation of S(t). One of the goals of survival analysis is the estimation of the survival function

$$S(t) = \exp\left(-\int_0^t \lambda(u)\,du\right).$$

For example, one of the yardsticks of a cancer therapy is the probability of surviving at least five years. In the engineering literature the survival function is called the reliability function (usually denoted R(t)), and a possible concern is the reliability of a component after 1000 hours.

Once we have the MLE, estimation of the survival function is easy under the exponential or Weibull model.

Exponential

$$\hat{S}(t) = e^{-\hat{\lambda}t},$$

Weibull

$$\hat{S}(t) = e^{-(\hat{\lambda}t)^{\hat{\alpha}}} = e^{-\hat{\gamma}t^{\hat{\alpha}}}.$$

Also, for any fixed t, $\hat{S}(t)$ is a function of $\hat{\lambda}$ or $(\hat{\gamma}, \hat{\alpha})$, so we can get an approximate distribution of $\hat{S}(t)$ by using the delta method. Alternatively, we can take a log log transformation that usually improves the convergence to normality.

Exponential

$$S(t) = e^{-\lambda t} ,$$

$$\log[-\log S(t)] = \log \lambda + \log t ,$$

$$\log[-\log \hat{S}(t)] = \log \hat{\lambda} + \log t ,$$

$$\hat{\mathrm{Var}}\{\log[-\log \hat{S}(t)]\} \cong \frac{1}{n_u} .$$

Weibull

$$S(t) = e^{-\gamma t^{\alpha}} ,$$

$$\log[-\log S(t)] = \log \gamma + \alpha \log t ,$$

$$\log[-\log \hat{S}(t)] = \log \hat{\gamma} + \hat{\alpha} \log t ,$$

$$\hat{\mathrm{Var}}\{\log[-\log \hat{S}(t)]\} \cong \frac{\mathrm{Var}(\hat{\gamma})}{\hat{\gamma}^2} + 2 \; \mathrm{Cov}(\hat{\gamma},\hat{\alpha}) \frac{\log t}{\hat{\gamma}}$$

$$+ \mathrm{Var}(\hat{\alpha})(\log t)^2 .$$

2.2. Linear Combinations of Order Statistics

In this section we consider only the Weibull distribution, but the ideas illustrated here can be generalized.

First, by reparametrizing and transforming, we can change the problem of estimating λ and α in the Weibull distribution to estimating location and scale parameters. Rewrite

$$P\{Y > t\} = e^{-(\lambda t)^{\alpha}} ,$$

$$= \exp\{-\exp[\alpha(\log \lambda + \log t)]\} ,$$

$$= \exp\left\{-\exp\left(\frac{\log t - \mu}{\sigma}\right)\right\} ,$$

where $\mu = -\log \lambda$ and $\sigma = 1/\alpha$. Then,

$$P\{\log Y > t\} = P\{Y > e^t\} ,$$

$$= \exp\left\{-\exp\left(\frac{t - \mu}{\sigma}\right)\right\} . \qquad (3)$$

From this we see that μ and σ are the location and scale parameters of the random variable $\log Y$. This is a useful observation since there is considerable statistical theory for estimating location and scale parameters.

Suppose we want to estimate the probability of survival for some fixed time t_0.

$$S(t_0) = P\{Y > t_0\} = \exp\left\{-\exp\left(\frac{\log t_0 - \mu}{\sigma}\right)\right\} .$$

Define $Y^0 = Y/t_0$ and $\mu_0 = \mu - \log t_0$. Then,

$$S(t_0) = P\{\log Y^0 > 0\} = \exp\left\{-\exp\left(\frac{-\mu_0}{\sigma}\right)\right\} ,$$

and μ_0 and σ are the location and scale parameters for $\log Y^0$. If we can construct a confidence interval for their ratio μ_0/σ, then by taking exponentials twice we would have a confidence interval for $S(t_0)$.

Johns and Lieberman use linear combination of order statistics:

$$\hat{\mu} = \sum_{i=1}^{n} a_i \log Y^0_{(i)} ,$$

$$\hat{\sigma} = \sum_{i=1}^{n} b_i \log Y^0_{(i)} \ ,$$

where $\sum_{i=1}^{n} a_i = 1$ and $\sum_{i=1}^{n} b_i = 0$ and
$a_1, \ldots, a_n, b_1, \ldots, b_n$ are chosen to satisfy an
asymptotic optimality criteria. This method is particularly well suited for Type II censoring where

$$a_{r+1} = \cdots = a_n = 0 \ ,$$

$$b_{r+1} = \cdots = b_n = 0 \ ,$$

so that the estimates are based only on the uncensored observations.

REFERENCE
Johns and Lieberman, Technometrics (1966).

Extreme Value Distributions. The function

$$G_1(x) = \exp[-\exp(-x)] \ , \quad -\infty < x < +\infty \ ,$$

is one of the three possible limiting extreme value
distributions. A limiting extreme value distribution is a distribution G for which there exists a
distribution function F such that if $X_1, \ldots, X_n$
are iid each with distribution F , then
$\max\{X_1, \ldots, X_n\}$, properly normalized, converges in
distribution to G. Another limiting extreme value
distribution is

$$G_2(x) = \begin{cases} \exp\{-(-x)^{\alpha}\} \ , & x < 0 \ , \\ 1 & , \quad x > 0 \ . \end{cases}$$

The upper tail of the Weibull distribution is the same as the lower tail of G_2, properly scaled, and the upper tail of the distribution of the log of a Weibull random variable (3) is the same as the lower tail of G_1. The d.f. G_2 arises as a normalized limit of

$$\max\{X_1, \ldots, X_n\} - x_0 \, ,$$

where x_0 is the upper truncation point for F [i.e., $F(x_0) = 1$, $F(x_0-) < 1$]. Since

$$-\max\{X_1 - x_0, \ldots, X_n - x_0\}$$
$$= \min\{x_0 - X_1, \ldots, x_0 - X_n\} \, ,$$

a Weibull random variable can be interpreted as the minimum (i.e., first failure) of a large number of potential failure times. The system fails with the occurrence of the first component failure.

2.3.   Other Estimators

The estimators in this section assume the exponential model with no censoring.

Bias-Corrected Estimators.   The method here is more important than the results. Suppose we estimate the survival probability with

$$\hat{S}(t) = e^{-\hat{\lambda}t} = e^{-t/\bar{T}} \, ,$$

where $\bar{T} = (1/n)\sum_{i=1}^{n} T_i$. Then,

$$E(\hat{S}(t)) \neq e^{-\lambda t} \, ,$$

so $\hat{S}(t)$ is a biased estimate. We can remove some of the bias by the delta method. If we denote $\theta = E(T)$,

$$e^{-t/\bar{T}} = e^{-t/\theta} + (\bar{T} - \theta) \frac{t}{\theta^2} e^{-t/\theta}$$

$$+ \frac{1}{2} (\bar{T} - \theta)^2 \left[\left(\frac{t}{\theta^2}\right)^2 - \frac{2t}{\theta^3}\right] e^{-t/\theta} + \cdots ,$$

$$E(e^{-t/\bar{T}}) = e^{-t/\theta} + 0 + \frac{1}{2} \frac{\theta^2}{n} \left[\left(\frac{t}{\theta^2}\right)^2 - \frac{2t}{\theta^3}\right] e^{-t/\theta}$$

$$+ \cdots ,$$

$$= \left[1 + \frac{1}{2n}\left(\frac{t^2}{\theta^2} - \frac{2t}{\theta}\right)\right] e^{-t/\theta} + \cdots .$$

Therefore,

$$\tilde{S}(t) = \frac{e^{-\hat{\lambda}t}}{1 + \frac{1}{2n} (t^2 \hat{\lambda}^2 - 2t \hat{\lambda})}$$

should be a less biased estimate of $S(t)$ than $\hat{S}(t)$. Also it usually turns out that $\tilde{S}(t)$ has smaller mean square error than $\hat{S}(t)$.

The jackknife estimate, which we discuss later, produces the same sort of bias correction.

Uniformly Minimum Variance Unbiased Estimators (UMVUEs). To get a UMVUE of $S(t)$, use the unbiased estimate

$$U = I(T_1 > t)$$

and the sufficient statistic

$$S = \sum_{i=1}^{n} T_i \; .$$

By the Rao-Blackwell theorem the UMVUE is $E(U|S)$, which in this case is

$$\tilde{S}(t) = E\{U|S = s\} = \left(1 - \frac{t}{s}\right)^{n-1} I(t < s) \; .$$

Bayesian Estimates. We only mention that Bayes estimates can be derived using gamma priors.

REFERENCES
Basu, Technometrics (1964), derives UMVUEs.
Zacks and Even, JASA (1966), compares mean square errors.
Gaver and Hoel, Technometrics (1970), look at estimators in the framework of sampling from a Poisson process.

3. REGRESSION MODELS

In medical applications the survival time may depend on the dose of medication or radiation, and in engineering applications the lifetime of a tube may depend on the temperature or other stress conditions.

Let Y denote the dependent variable and x denote the independent variable. Feigl and Zelen propose two models for the exponential distribution.

(i) The linear model

$$E(T) = \alpha + \beta x \; .$$

To get estimates, use maximum likelihood. The drawback with this model is the possibility of obtaining a negative estimate of $E(T)$ when $\hat{\beta}$ is negative.

(ii) <u>The log-linear model</u>

$$E(T) = \alpha e^{\beta x} \; ,$$

$$\log E(T) = \log \alpha + \beta x \; .$$

Again, use maximum likelihood. This model is the precursor to the Cox proportional hazards model.

Also, see Note (i) in Chapter 6, Section 2.2.

REFERENCES

Feigl and Zelen, <u>Biometrics</u> (1965), discuss the uncensored case for both the linear and log-linear models.

Zippin and Armitage, <u>Biometrics</u> (1966), discuss the censored case for the linear model.

Glasser, <u>JASA</u> (1967), discusses the censored case for the log-linear model.

Zippin and Lamborn, Stanford Univ. Tech. Report No. 20 (1969), discuss the censored case for the log-linear model and goodness of fit tests.

Mantel and Myers, <u>JASA</u> (1971), discuss the censored case for the multiple linear model.

4. MODELS WITH SURVIVING FRACTIONS

4.1. Single Sample

Let

$$p = P\{death\} \quad \text{and} \quad 1 - p = P\{survival\} \; ,$$

where the latter probability is called the <u>surviving fraction</u>. Assume

$$P\{T \leq t \,|\, death\} = 1 - e^{-\lambda t} \; .$$

Then the likelihood is

$$L(y,\delta) = \begin{cases} p\lambda e^{-\lambda y} & \text{if} \quad \delta = 1 \quad \text{(uncensored)} , \\\\ (1-p) + pe^{-\lambda y} & \text{if} \quad \delta = 0 \quad \text{(censored)} . \end{cases}$$

To get estimates, use maximum likelihood.

Models with surviving fractions are sometimes used for short-term experiments where one does not hypothesize that the survival function $S(t)$ necessarily approaches zero. Instead, $S(t)$ may have the following form:

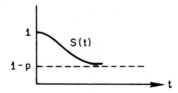

4.2. Regression

Assume

$$p(x) \equiv P\{death|x\} = \frac{e^{\alpha+\beta x}}{1 + e^{\alpha+\beta x}} .$$

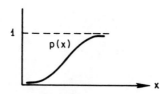

This is the logistic function. Also, let

$$P\{T \leq t \mid \text{death}\} = 1 - e^{-\lambda t} \ .$$

The likelihood is

$$L(y,\delta,x) = \begin{cases} p(x) \ \lambda e^{-\lambda y} & \text{if} \quad \delta = 1 \quad \text{(uncensored)}, \\[2ex] 1 - p(x) + p(x) e^{-\lambda y} & \text{if} \quad \delta = 0 \\ & \qquad\qquad \text{(censored)} \ . \end{cases}$$

To get estimates, use maximum likelihood.

REFERENCE
Farewell, _Biometrika_ (1977).

THREE

NONPARAMETRIC METHODS: ONE SAMPLE

1. LIFE TABLES

The classical method of estimating $S(t)$ in epidemiology and actuarial science is the actuarial method discussed below. It depends on the <u>life table</u>.

Let time be partitioned into a fixed sequence of intervals $I_1, \ldots, I_k$. These intervals are almost always, but not necessarily, of equal lengths, and for human populations the length of each interval is usually one year.

For a life table let

n_i = # alive at the beginning of I_i ,

d_i = # died during I_i ,

ℓ_i = # lost to follow-up during I_i ,

w_i = # withdrew during I_i ,

p_i = P{surviving through I_i|alive at

beginning of I_i} ,

$q_i = 1 - p_i$.

Table 1 is an example of a life table. $I_1, I_2, \ldots, I_5$ each has length one year. Column (2) contains n_i , (3) contains d_i , (4) contains ℓ_i , and (5) contains w_i. We want to estimate $S(5$ years).

1.1. Reduced Sample Method

To estimate $S(\tau_k)$, use only those subjects who are at risk during $(0, \tau_k]$, the entire interval of interest. Let

$$n = n_1 - \sum_{i=1}^{k} \ell_i - \sum_{i=1}^{k} w_i ,$$

$$d = \sum_{i=1}^{k} d_i ,$$

$$\hat{S}(\tau_k) = 1 - \frac{d}{n} .$$

For the example in Table 1,

TABLE 1. COMPUTATION OF THE 5-YEAR SURVIVAL RATE

Years After Diagnosis (1)	Alive at Beginning of Interval (2)	Died During Interval (3)	Lost to Followup During Interval (4)	Withdrawn Alive During Interval (5)
0-1	126	47	4	15
1-2	60	5	6	11
2-3	38	2	—	15
3-4	21	2	2	7
4-5	10	—	—	6

41

TABLE 1 (Continued)

Effective Number Exposed to the Risk of Dying $(2) - \frac{1}{2}[(4) + (5)]$	Proportion Dying $(3)/(6)$	Proportion Surviving $1 - (7)$	Cumulative Proportion Surviving From Diagnosis Through End of Interval $\Pi_1^k (8)_i$
(6)	(7)	(8)	(9)
116.5	0.40	0.60	0.60
51.5	0.10	0.90	0.54
30.5	0.07	0.93	0.50
16.5	0.12	0.88	0.44
7.0	0.00	1.00	0.44

REFERENCE: Cutler and Ederer, J. Chronic Dis. (1958).

$$n = 126 - 12 - 54 = 60 \; ,$$

$$d = 56 \; ,$$

$$\hat{S}(5 \text{ years}) = 1 - \frac{56}{60} = 0.078 \; .$$

The drawback with the reduced sample method is that it ignores the information that is contained in ℓ_i and w_i. It is a biased (downward) estimate of $S(t)$.

1.2. Actuarial Method

We can break up the survival probability $S(\tau_k)$ into a product of probabilities:

$$S(\tau_k) = P\{T > \tau_k\} \; ,$$

$$= P\{T > \tau_1\} \; P\{T > \tau_2 | T > \tau_1\} \; \cdots$$

$$P\{T > \tau_k | T > \tau_{k-1}\} \; ,$$

$$= P_1 \cdot P_2 \; \cdots \; P_k \; ,$$

where

$$P_i = P\{T > \tau_i | T > \tau_{i-1}\} \; .$$

The actuarial method gives an estimate for each P_i separately and then multiplies the estimates together to estimate $S(\tau_k)$.

For an estimate of P_i , we could use $1 - d_i/n_i$, if there were no losses or withdrawals in I_i. However, with ℓ_i and w_i nonzero, we assume that, on the average, those individuals who became lost or withdrawn during I_i were at risk for half the interval. Therefore, define the <u>effective sample size</u>

$$n_i' = n_i - \frac{1}{2}(\ell_i + w_i) \;,$$

and

$$\hat{q}_i = \frac{d_i}{n_i'} \;,$$

$$\hat{p}_i = 1 - \hat{q}_i \;.$$

The <u>actuarial estimate</u> is

$$\hat{S}(\tau_k) = \prod_{i=1}^{k} \hat{p}_i \;.$$

In Table 1, column (6) contains n_i' , (7) contains $\hat{q}_i$, (8) contains $\hat{p}_i$, and in column (9) we see

$$\hat{S}(5 \text{ years}) = 0.44 \;.$$

There has been work on finding an improved substitute for the effective sample size, but if a finer estimate of $S(t)$ is required, the product-limit estimator of Kaplan and Meier is the approach to take.

1.3. Variance of $\hat{S}(\tau_k)$

To estimate the variance of $\hat{S}(\tau_k)$, consider

$$\log \hat{S}(\tau_k) = \sum_{i=1}^{k} \log \hat{p}_i \;.$$

Assuming $n_i' \hat{p}_i \approx \text{Binomial}(n_i', p_i)$, the delta method implies

$$\text{Var}(\log \hat{p}_i) \cong \text{Var}(\hat{p}_i)\left(\frac{d}{dp_i}(\log p_i)\right)^2$$

$$\cong \frac{p_i q_i}{n_i'} \cdot \frac{1}{p_i^2} = \frac{q_i}{n_i' p_i} \,,$$

and assuming $\log \hat{p}_1, \ldots, \log \hat{p}_k$ are independent,

$$\text{Var}[\log \hat{S}(\tau_k)] \cong \sum_{i=1}^{k} \frac{q_i}{n_i' p_i} \,,$$

$$\hat{\text{Var}}[\log \hat{S}(\tau_k)] = \sum_{i=1}^{k} \frac{\hat{q}_i}{n_i' \hat{p}_i} = \sum_{i=1}^{k} \frac{d_i}{n_i'(n_i' - d_i)} \,.$$

Now using the delta method again,

$$\hat{\text{Var}}(\hat{S}(\tau_k)) = \hat{S}^2(\tau_k) \sum_{i=1}^{k} \frac{d_i}{n_i'(n_i' - d_i)} \,,$$

which is called Greenwood's formula.

1.4. Types of Life Tables

Table 1 is an example of a cohort life table. A cohort is a group of people who are followed throughout the course of the study. The people at risk at the beginning of the interval I_i are those people who survived (not dead, lost, or withdrawn) the previous interval I_{i-1}.

Another type of life table is the current life table. In a current life table a group of people with age τ_{i-1} are considered to be at risk at the beginning of the interval $I_i = (\tau_{i-1}, \tau_i]$, and this group of people is completely different from those at risk in the previous interval I_{i-1}. Typically, different age groups in the population are followed at the same time.

REFERENCES
Berkson and Gage, <u>Proc. Staff Meet. Mayo Clin.</u>
   (1950).
Cutler and Ederer, <u>J. Chronic Dis.</u> (1958).
Elveback, <u>JASA</u> (1958).
Chiang, <u>Stochastic Processes in Biostatistics</u>
   (1968), Chapter 9.
Breslow and Crowley, <u>Ann. Stat.</u> (1974).

2. PRODUCT-LIMIT (KAPLAN-MEIER) ESTIMATOR

The <u>product-limit (PL) estimator</u> is similar to the actuarial estimator except the lengths of the intervals I_i are variable. In fact, let τ_i , the right endpoint of I_i , be the i-th ordered censored or uncensored observation.

$$\left(\overset{I_1}{\rule{0pt}{0pt}} \right]\left[\overset{I_2}{\rule{0pt}{0pt}} \right]\left[\overset{I_3}{\rule{0pt}{0pt}} \right]\left[\overset{I_4}{\rule{0pt}{0pt}} \right] \quad \cdots \quad \left(\overset{I_n}{\rule{0pt}{0pt}} \right]$$

```
|----X----0----0----X----0----X----> t
   τ_1  τ_2  τ_3  τ_4   τ_{n-1}  τ_n
```

0 = censored and X = uncensored.

Recall that we observe the pairs (Y_1, δ_1), $\cdots$, (Y_n, δ_n). For now, assume no ties. Let $Y_{(1)} < Y_{(2)} < \cdots < Y_{(n)}$ be the order statistics of $Y_1, Y_2, \cdots, Y_n$, and with an abuse of notation, define $\delta_{(i)}$ to be the value of δ associated with $Y_{(i)}$, that is, $\delta_{(i)} = \delta_j$ when $Y_{(i)} = Y_j$. Note that $\delta_{(1)}, \ldots, \delta_{(n)}$ are not ordered. Let $\mathcal{R}(t)$ denote the <u>risk set</u> at time t, which is the set of subjects still alive at time $t-$, and let

$$n_i = \text{\# in } \mathcal{R}(Y_{(i)}) = \text{\# alive at time } Y_{(i)}^{-},$$

$$d_i = \text{\# died at time } Y_{(i)},$$

$$p_i = P\{\text{surviving through } I_i | \text{alive at beginning}$$

$$\text{of } I_i\},$$

$$= P\{T > \tau_i | T > \tau_{i-1}\},$$

$$q_i = 1 - p_i.$$

From the estimates

$$\hat{q}_i = \frac{d_i}{n_i},$$

$$\hat{p}_i = 1 - \hat{q}_i = \begin{cases} 1 - \dfrac{1}{n_i} & \text{if } \delta_{(i)} = 1 \quad (\text{uncensored}), \\[2mm] 1 & \text{if } \delta_{(i)} = 0 \quad (\text{censored}), \end{cases}$$

the PL estimate when no ties are present is

$$\hat{S}(t) = \prod_{y_{(i)} \leq t} \hat{p}_i = \prod_{u: y_{(i)} \leq t} \left(1 - \frac{1}{n_i}\right),$$

$$= \prod_{y_{(i)} \leq t} \left(1 - \frac{1}{n_i}\right)^{\delta_{(i)}},$$

$$= \prod_{y_{(i)} \leq t} \left(1 - \frac{1}{n - i + 1}\right)^{\delta_{(i)}},$$

$$= \prod_{y_{(i)} \le t} \left(\frac{n - i}{n - i + 1} \right)^{\delta(i)} .$$

REFERENCE
Kaplan and Meier, JASA (1958).

NOTES

(i) For tied uncensored observations, suppose just before time t , there are m individuals alive, and at time t, d deaths occur. Split the time of the d deaths infinitesimally so that the factor for the d deaths in the product-limit estimator is

$$\left(1 - \frac{1}{m} \right)\left(1 - \frac{1}{m - 1} \right) \cdots \left(1 - \frac{1}{m - d + 1} \right)$$

$$= \left(\frac{m - 1}{m} \right)\left(\frac{m - 2}{m - 1} \right) \cdots \left(\frac{m - d}{m - d + 1} \right)$$

$$= \frac{m - d}{m} = 1 - \frac{d}{m} .$$

(ii) If censored and uncensored observations are tied, consider the uncensored observations to occur just before the censored observations.

(iii) If the last (ordered) observation $y_{(n)}$ is censored, then for $\hat{S}(t)$ as defined above

$$\lim_{t \to \infty} \hat{S}(t) > 0 .$$

Sometimes it is preferable to redefine $\hat{S}(t) = 0$ for $t \ge y_{(n)}$ or to think of it as being undefined for $t \ge y_{(n)}$ if $\delta_{(n)} = 0.$

From Notes (i) and (ii), by letting

$$y'_{(1)} < y'_{(2)} < \cdots < y'_{(r)}$$

denote the distinct survival times and

$$\delta'_{(j)} = \begin{cases} 1 & \text{if the observations at time } y'_{(j)} \\ & \text{are uncensored,} \\ 0 & \text{if censored,} \end{cases}$$

$$n_j = \# \text{ in } \mathcal{R}(y'_{(j)}),$$

$$d_j = \# \text{ died at time } y'_{(j)},$$

the PL estimate allowing for ties is

$$\hat{S}(t) = \prod_{u:y'_{(j)} \le t} \left(1 - \frac{d_j}{n_j}\right) = \prod_{y'_{(j)} \le t} \left(1 - \frac{d_j}{n_j}\right)^{\delta'_{(j)}}.$$

Example. AML Maintenance Study. A clinical trial to evaluate the efficacy of maintenance chemotherapy for acute myelogenous leukemia (AML) was conducted by Embury et al. at Stanford University. After reaching a state of remission through treatment by chemotherapy, the patients who entered the study were randomized into two groups. The first group received maintenance chemotherapy; the second or control group did not. The objective of the trial was to see if maintenance chemotherapy prolonged the time until relapse, that is, increased the length of remission.

For a preliminary analysis during the course of the trial the data (on 10/74) were as follows:

Length of complete remission (in weeks)

Maintained group

9, 13, 13+, 18, 23, 28+, 31, 34, 45+, 48, 161+.

Nonmaintained group

5, 5, 8, 8, 12, 16+, 23, 27, 30, 33, 43, 45.

The Kaplan–Meier PL estimator for the maintained group is computed as follows:

$$\hat{S}(0) = 1 ,$$

$$\hat{S}(9) = \hat{S}(0) \times \frac{10}{11} = .91 ,$$

$$\hat{S}(13) = \hat{S}(9) \times \frac{9}{10} = .82 ,$$

$$\hat{S}(18) = \hat{S}(13) \times \frac{7}{8} = .72 ,$$

$$\hat{S}(23) = \hat{S}(18) \times \frac{6}{7} = .61 ,$$

$$\hat{S}(31) = \hat{S}(23) \times \frac{4}{5} = .49 ,$$

$$\hat{S}(34) = \hat{S}(31) \times \frac{3}{4} = .37 ,$$

$$\hat{S}(48) = \hat{S}(34) \times \frac{1}{2} = .18 .$$

Figure 3 exhibits the PL estimators for the maintained and nonmaintained groups.

REFERENCE
Embury et al., West. J. Med. (1977).

Variance of $\hat{S}(t)$. Using the same arguments as for the variance of the actuarial estimate, we can get in the case of no ties

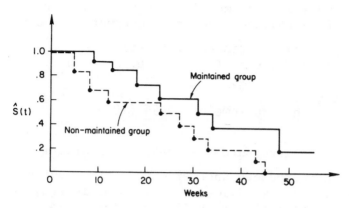

Figure 3. Estimated survival functions for AML maintenance study.

$$\hat{\text{Var}}(\hat{S}(t)) = \hat{S}^2(t) \sum_{y_{(i)} \leq t} \frac{\hat{q}_i}{n_i \hat{p}_i} \, ,$$

$$= \hat{S}^2(t) \sum_{y_{(i)} \leq t} \frac{\delta_{(i)}}{(n - i)(n - i + 1)} \, .$$

With ties present

$$\hat{\text{Var}}(\hat{S}(t)) = \hat{S}^2(t) \sum_{y'_{(i)} \leq t} \frac{\delta'_{(j)} \, d_j}{n_j(n_j - d_j)} \, .$$

These formulas are referred to as <u>Greenwood's formula</u> as well.

The justification for these formulas is not as clear as in the case of life tables because the number of terms in the product is random and there is more dependence between the terms. However, they can be justified as approximations to the asymptotic variance of $\hat{S}(t)$ to be given later.

Thomas and Grunkemeier study three different methods of confidence interval construction.   One uses
the approximate variance $\text{Var}(\hat{S}(t))$.   Also, see the
comments at the end of Chapter 6, Section 1.4.

REFERENCE
Thomas and Grunkemeier, JASA (1975).

2.1.   Redistribute-to-the-Right Algorithm

Efron introduced another method of computing the
PL estimator.   We illustrate with the leukemia (AML)
example.   Plot the   (n = 11)   survival times:

$$9 \quad 13 \; 13+ \qquad 18 \qquad 23 \qquad 28+ \quad 31 \cdots 161+$$

The ordinary estimate of   S(t)   assuming no censoring
puts mass   1/11   at each observed time.   Consider the
first censored time   13+.   Since a death did not occur at   13+   but somewhere to the right of it, it
seems reasonable to redistribute   1/11 , the mass at
13+ , equally among all observed times to the right
of   13+.   Therefore, add   (1/8)(1/11)   to the mass at
18, 23, 28+, ...   Now consider the next censored
time   28+ ; redistribute   1/11 + (1/8)(1/11) , the
mass at   28+ , among all observed times to the right
of   28+.   Treating the other censored times similarly
results in the PL estimator given on the next page.

REFERENCE
Efron, Proc. Fifth Berkeley Symp. IV (1967),
    pp. 831-853.

2.2.   Self-Consistency

For simplicity, we will assume no ties.   An estimator $\hat{SC}(t)$ is self-consistent if

y(i)	Mass at Start	Mass After First Redistribution	Mass After Second Redistribution	Mass After Third Redistribution	$\hat{S}(y_{(i)})$
9	1/11=.09	.09	.09	.09	.91
13	.09	.09	.09	.09	.82
13+	.09	0	0	0	
18	...	.09+(1/8)(.09)=.10	.10	.10	.72
23	...	.10	.10	.10	.61
28+		.10	0	0	
31		...	.10+(1/5)(.10)=.12	.12	.49
34			.12	.12	.37
45+			.12	0	
48			...	.12+(1/2)(.12)=.18	.18
161+				.18	

53

$$\hat{SC}(t) = \frac{1}{n}\left[\sum_{i=1}^{n} 1 \cdot I(y_{(i)} > t)\right.$$

$$+ \sum_{i=1}^{n} 0 \cdot I(y_{(i)} \leq t, \ \delta_{(i)} = 1) \qquad (4)$$

$$+ \left. \sum_{i=1}^{n} \frac{\hat{SC}(t)}{\hat{SC}(y_{(i)})} \ I(y_{(i)} \leq t, \ \delta_{(i)} = 0)\right] \ ,$$

where $\hat{SC}(t)/\hat{SC}(y_{(i)})$ estimates the conditional probability of surviving beyond t given alive at $y_{(i)}$.
Notice that (4) is equivalent to

$$\hat{SC}(t) = \frac{1}{n}\left[N_y(t) + \sum_{y_{(i)} \leq t} (1 - \delta_{(i)}) \frac{\hat{SC}(t)}{\hat{SC}(y_{(i)})}\right], \ (5)$$

where

$$N_y(t) = \#(y_i > t) \ .$$

(The notation "#(__)" denotes "the number satisfying
__.")
  The PL estimator is the unique self-consistent
estimator for $t < y_{(n)}$. The proof proceeds as
follows.
  From (5), a self-consistent estimator satisfies

$$\hat{SC}(t) = \frac{N_y(t)}{n - \sum_{y_{(i)} \le t} \left(\frac{1 - \delta_{(i)}}{\hat{SC}(y_{(i)})}\right)} \ ,$$

$$= \begin{cases} 1 & \text{if } t < y_{(1)} \ , \quad (6) \\[2ex] \dfrac{N_y(t)}{n - \sum\limits_{i=1}^{k} \left(\dfrac{1 - \delta_{(i)}}{\hat{SC}(y_{(i)})}\right)} & \text{if } y_{(k)} \le t < y_{(k+1)} \ , \\[2ex] & k = 1, 2, \ldots, n-1 \ . \end{cases}$$

We want to show that if $\hat{SC}(t)$ satisfies (6), then $\hat{SC}(t)$ coincides with the PL estimator $\hat{S}(t)$. First notice that

$$\hat{S}(t) = 1 = \hat{SC}(t) \quad \text{if } t < y_{(1)} \ .$$

Also, $\hat{S}(t)$ and $\hat{SC}(t)$ are constant on $[y_{(k)}, y_{(k+1)})$ for $k = 1, \ldots, n - 1$. Therefore, we need only show that the jump at $y_{(k)}$ of $\hat{SC}(t)$ is the same as the jump of $\hat{S}(t)$.

(i) If $\delta_{(k)} = 0$, (6) implies

$$N_y(y_{(k)}-) - 1 = N_y(y_{(k)}) \ ,$$

$$= \hat{SC}(y_{(k)}) \left[n - \sum_{i=1}^{k} \left(\frac{1 - \delta_{(i)}}{\hat{SC}(y_{(i)})}\right)\right] \ ,$$

$$= \hat{SC}(y_{(k)}) \left[n - \sum_{i=1}^{k-1} \left(\frac{1 - \delta_{(i)}}{\hat{SC}(y_{(i)})}\right)\right] - 1 \ ,$$

$$= \hat{SC}(y_{(k)}) \left[\frac{N_y(y_{(k)}-)}{\hat{SC}(y_{(k)}-)} \right] - 1 \, ,$$

which implies

$$\hat{SC}(y_{(k)}) = \hat{SC}(y_{(k)}-) \, .$$

Thus, $\hat{SC}(t)$ has no jump at $y_{(k)}$ when $\delta_{(k)} = 0$, and therefore agrees with $\hat{S}(t)$ at $t = y_{(k)}$.

(ii) If $\delta_{(k)} = 1$, (6) implies

$$\hat{SC}(y_{(k)}) = \frac{N_y(y_{(k)})}{n - \sum\limits_{i=1}^{k} \left(\frac{1 - \delta_{(i)}}{\hat{SC}(y_{(i)})} \right)} \, ,$$

$$= \frac{N_y(y_{(k)})}{N_y(y_{(k)}-)} \frac{N_y(y_{(k)}-)}{n - \sum\limits_{i=1}^{k-1} \left(\frac{1 - \delta_{(i)}}{\hat{SC}(y_{(i)})} \right)} \, ,$$

$$= \frac{n - k}{n - k + 1} \hat{SC}(y_{(k)}-) \, ,$$

so $\hat{SC}(t)$ has a jump at $y_{(k)}$ if $\delta_{(k)} = 1$ with $\hat{SC}(y_{(k)})/\hat{SC}(y_{(k)}-) = (n - k)/(n - k + 1)$, again agreeing with $\hat{S}(t)$ at $t = y_{(k)}$.

Self-consistency algorithm. Consider the naive estimator

$$\hat{S}^0(t) = \frac{N_y(t)}{n} \, .$$

This estimator can be improved by iteration using

$$\hat{s}^{(j+1)}(t) = \frac{1}{n}\left[N_y(t) + \sum_{y_{(i)} \leq t} (1-\delta_{(i)})\frac{\hat{s}^{(j)}(t)}{\hat{s}^{(j)}(y_{(i)})}\right].$$

In fact, $\hat{s}^{(j)}(t)$ converges monotonically in a finite number of steps to the PL estimator. This computational algorithm can be useful in more general censoring problems.

REFERENCES
Efron, Proc. Fifth Berkeley Symp. IV (1967).
Turnbull, JASA (1974).
_____, JRSS B (1976).

2.3. Generalized Maximum Likelihood Estimator

In the usual setup, we assume that our observation X has a probability measure P_θ that satisfies

$$dP_\theta(\underset{\sim}{x}) = f_\theta(\underset{\sim}{x})\ d\mu(\underset{\sim}{x}) ,$$

where $\mu(\underset{\sim}{x})$ is a dominating measure for the class $\{P_\theta\}$. Getting the maximum likelihood estimator of θ involves maximizing the likelihood

$$L(\theta) = f_\theta(\underset{\sim}{x}) .$$

In our case, we assume that our observation has a probability measure P_F that depends on the unknown distribution function F. The class $\{P_F\}$ has no dominating measure so we need a more general definition of maximum likelihood.

Kiefer and Wolfowitz suggest the following definition. Let $\wp = \{P\}$ be a class of probability

measures. For the elements P_1 and P_2 in $\mathcal{P}$,
define

$$f(\underset{\sim}{x};\ P_1,\ P_2) = \frac{dP_1(\underset{\sim}{x})}{d(P_1 + P_2)}\ ,$$

the Radon-Nikodym derivative of P_1 with respect to
$P_1 + P_2$. Define the probability measure $\hat{P}$ to be a
generalized maximum likelihood estimator (GMLE) if

$$f(\underset{\sim}{x};\ \hat{P},\ P) \geq f(\underset{\sim}{x};\ P,\ \hat{P}) \tag{7}$$

for any element $P \in \mathcal{P}$.

   This definition of the GMLE includes the definition of the usual MLE.

   The Kaplan-Meier PL estimator gives the GMLE of F.
The proof proceeds as follows. For convenience
assume no ties.

   If a probability measure $\hat{P}$ gives positive probability to $\underset{\sim}{x}$, then $f(x;P,\hat{P}) = 0$ unless P also
gives positive probability to $\underset{\sim}{x}$. Thus, to check (7)
for $P \in \mathcal{P}$ it is sufficient to check it for those P
with $P\{\underset{\sim}{x}\} > 0$ and in this case (7) reduces to

$$\hat{P}\{\underset{\sim}{x}\} \geq P\{\underset{\sim}{x}\}\ . \tag{8}$$

   Since $\hat{S}$ puts positive mass on the point
$\underset{\sim}{x} = ((y_1,\ \delta_1),\ \ldots,\ (y_n,\ \delta_n))$, we need only consider
probability measures P which put positive mass on
this point and show that $\hat{S}$ maximizes
$P\{((y_1,\ \delta_1),\ \ldots,\ (y_n,\ \delta_n))\}$. For any such P

$$L = P\{(y_1,\ \delta_1),\ \ldots,\ (y_n,\ \delta_n)\}\ ,$$

$$= \prod_{i=1}^{n} P\{T = y_{(i)}\}^{\delta(i)} \, P\{T > y_{(i)}\}^{1-\delta(i)} \ .$$

Let P assign probability p_i to the half-open interval $[y_{(i)}, y_{(i+1)})$, where $y_{(n+1)} = +\infty$. For fixed $p_1, \ldots, p_n$ the likelihood L is maximized by setting $P\{T = y_{(i)}\} = p_i$ if $\delta_{(i)} = 1$. If $\delta_{(i)} = 0$, then L is maximized by setting $P\{y_{(i)} < T < y_{(i+1)}\} = p_i$. Thus, for fixed $p_1, \ldots, p_n$ the maximum value for L is

$$\prod_{i=1}^{n} p_i^{\delta(i)} \left(\sum_{j=i}^{n} p_j \right)^{1-\delta(i)} \ . \tag{9}$$

From Problem 8 we see that (9) is maximized by

$$\hat{p}_i = \prod_{j=1}^{i-1} \left(1 - \frac{\delta(j)}{n - j + 1} \right) \frac{\delta(i)}{n - i + 1} \ .$$

This corresponds to $\hat{S}$. The argument for ties is identical.

REFERENCES
Kiefer and Wolfowitz, Ann. Math. Stat. (1956).
Kaplan and Meier, JASA (1958).
Johansen, Scand. J. Stat. (1978).

2.4. Consistency

Recall

$$S(t) = S_T(t) = P\{T > t\} = 1 - F(t) \ ,$$

and define S^* by

$$S^*(t) = S_Y(t) = P\{Y > t\} = 1 - H(t) \ ,$$

$$= [1 - F(t)][1 - G(t)] \ .$$

Define the <u>subsurvival functions</u>

$$S_u^*(t) = P\{Y > t, \ \delta = 1\} = \int_t^\infty [1 - G(u)] \ dF(u) \ ,$$

$$S_c^*(t) = P\{Y > t, \ \delta = 0\} = \int_t^\infty [1 - F(u)] \ dG(u) \ .$$

Then,

$$S^*(t) = S_u^*(t) + S_c^*(t) \ .$$

We will show that $S(t)$ can be expressed as a function of $S_u^*(t)$ and $S_c^*(t)$.

(i) Suppose $S_u^*(t)$ is continuous.

$$\int_0^t \frac{dS_u^*(u)}{S_u^*(u) + S_c^*(u)} = \int_0^t \frac{-[1 - G(u)] \ dF(u)}{[1 - F(u)][1 - G(u)]}$$

$$= \int_0^t \frac{-dF(u)}{1 - F(u)} = \log[1 - F(u)] \Big|_0^t$$

$$= \log S(t) \ .$$

Therefore,

$$S(t) = \exp\left[\int_0^t \frac{dS_u^*(u)}{S_u^*(u) + S_c^*(u)}\right].$$

(ii) Suppose S_u^* has a jump at t, but S_c^* is continuous at t.

$$\log \frac{S_u^*(t+) + S_c^*(t+)}{S_u^*(t-) + S_c^*(t-)} = \log \frac{[1 - F(t+)][1 - G(t+)]}{[1 - F(t-)][1 - G(t-)]},$$

$$= \log \frac{[1 - F(t+)]}{[1 - F(t-)]} = \log \frac{S(t+)}{S(t-)}.$$

(The second equality follows from the fact that S_c^* is continuous at t so $G(t+) = G(t-)$.) Therefore,

$$S(t+) = S(t-) \exp\left\{\log\left[\frac{S_u^*(t+) + S_c^*(t+)}{S_u^*(t-) + S_c^*(t-)}\right]\right\}.$$

If the underlying distributions F and G have no common jumps, then from (i) and (ii)

$$S(t) = \exp\left\{c\int_0^t \frac{dS_u^*(u)}{S_u^*(u) + S_c^*(u)}\right.$$

$$\left. + d \sum_{u \leq t} \log\left[\frac{S_u^*(u+) + S_c^*(u+)}{S_u^*(u-) + S_c^*(u-)}\right]\right\}, \tag{10}$$

where $c\!\int$ denotes integration over the continuity intervals of S_u^* and $d\sum$ denotes summation over the discrete jumps of S_u^*. Expression (10), called Peterson's representation, shows that $S(t)$ can be represented as a function of S_u^*, S_c^*, and t, that is,

$$S(t) = \Psi(S_u^*, S_c^*; t) .$$

Peterson's representation gives us a proof that the PL estimator $\hat{S}(t)$ is consistent. The proof proceeds as follows.

Define the empirical subsurvival functions

$$\hat{S}_u^*(t) = \frac{1}{n} \sum_{i=1}^{n} I(y_i > t, \delta_i = 1) ,$$

$$\hat{S}_c^*(t) = \frac{1}{n} \sum_{i=1}^{n} I(y_i > t, \delta_i = 0) .$$

It can be seen that the PL estimator is

$$\hat{S}(t) = \Psi(\hat{S}_u^*, \hat{S}_c^*; t) ,$$

provided any ties between uncensored and censored observations are interpreted as uncensored observations preceding censored. Notice that since $\hat{S}_u^*$ is discrete, $\Psi(\hat{S}_u^*, \hat{S}_c^*; t)$ involves only the summation over the discrete jumps of $\hat{S}_u^*$.

By the Glivenko-Cantelli theorem,

$$\hat{S}_u^*(t) \xrightarrow{\text{a.s.}} S_u^*(t) \;,$$

$$\hat{S}_c^*(t) \xrightarrow{\text{a.s.}} S_c^*(t) \;, \quad \text{uniformly in} \quad t.$$

(The notation "$\xrightarrow{\text{a.s.}}$" denotes "converges almost surely to.") Also, Ψ is a continuous function of S_u^*, S_c^* in the sup norm. That is, if

$$\| S_u^* - S_u^{**} \| = \sup_t \, | S_u^*(t) - S_u^{**}(t) | \to 0$$

and

$$\| S_c^* - S_c^{**} \| \to 0 \;,$$

then

$$\Psi(S_u^*, \, S_c^*; \, t) \to \Psi(S_u^{**}, \, S_c^{**}; \, t) \;.$$

Therefore,

$$\hat{S}(t) = \Psi(\hat{S}_u^*, \, \hat{S}_c^*; \, t) \xrightarrow{\text{a.s.}} \Psi(S_u^*, \, S_c^*; \, t) = S(t) \;.$$

REFERENCE
Peterson, JASA (1977).

2.5. Asymptotic Normality

We will show later that if F and G are continuous on $[0, T]$ and $F(T) < 1$, then

$$Z_n(t) = \sqrt{n} \; [\hat{S}(t) - S(t)] \xrightarrow{W} Z(t) \quad \text{as} \quad n \to \infty \;,$$

where $Z(t)$ is a Gaussian process with moments

$$E(Z(t)) = 0 \ ,$$

$$\text{Cov}(Z(t_1), \ Z(t_2)) = S(t_1) \ S(t_2)$$

$$\times \int_0^{t_1 \wedge t_2} \frac{dF_u(u)}{[1 - H(u)]^2} \ ,$$

$$= S(t_1) \ S(t_2)$$

$$\times \int_0^{t_1 \wedge t_2} \frac{dF(u)}{[1 - F(u)][1 - H(u)]} \ ,$$

where

$$F_u(t) = P\{T \leq t, \ \delta = 1\} = \int_0^t [1 - G(u)] \ dF(u) \ ,$$

$$1 - H(u) = [1 - F(u)][1 - G(u)] \ .$$

The proof involves hazard functions which are to be discussed in the next section.

We remark that $Z_n(t)$ converges weakly ($\overset{w}{\rightarrow}$) to the Gaussian process $Z(t)$ means that for any $t_1, \ \ldots, \ t_k, \ Z_n(t_1), \ \ldots, \ Z_n(t_k)$ have an asymptotic multivariate normal distribution and the sequence of probability measures for Z_n is tight so that $f(Z_n)$ converges in distribution to $f(Z)$ for any function f continuous in the sup norm.

As a particular case of the above result,

$$\hat{S}(t) \overset{a}{\sim} N\left(S(t), \ \frac{S^2(t)}{n} \int_0^t \frac{dF_u(u)}{[1 - H(u)]^2}\right) \ .$$

We can obtain an approximation for the asymptotic variance of $\hat{S}(t)$. Because $F_u(t) = P\{T \leq t, \ \delta = 1\}$

and $H(t) = P\{Y \leq t\}$, let (assuming no ties)

$$d\hat{F}_u(y_{(i)}) = \frac{\delta_{(i)}}{n} \, ,$$

$$1 - \hat{H}(y_{(i)}) = 1 - \frac{i}{n} = \frac{n-i}{n} \, ,$$

$$1 - \hat{H}(y_{(i)}-) = 1 - \frac{i-1}{n} = \frac{n-i+1}{n} \, .$$

Replacement of $(1 - H(u))^2$ by $(1-H(u))(1-H(u-))$ in the asymptotic variance and substitution of the above estimates gives

$$A\hat{V}ar(\hat{S}(t)) = \frac{\hat{S}^2(t)}{n} \sum_{y_{(i)} \leq t} \frac{\delta_{(i)}/n}{[(n-i)/n][(n-i+1)/n]} ,$$

$$= \hat{S}^2(t) \sum_{y_{(i)} \leq t} \frac{\delta_{(i)}}{(n-i)(n-i+1)} \, ,$$

which is precisely Greenwood's formula. ("AVar" denotes "asymptotic variance.")

REFERENCES
Billingsley, Convergence of Probability Measures
    (1968), for weak convergence.
Breslow and Crowley, Ann. Stat. (1974).

3. HAZARD FUNCTION ESTIMATORS

Recall that the hazard function is

$$\lambda(t) = \frac{f(t)}{1 - F(t)} \, .$$

Estimating $\lambda(t)$ is essentially equivalent to the

difficult problem of estimating a density.   An easier
problem is estimating the <u>cumulative hazard function</u>:

$$\Lambda(t) = \int_0^t \lambda(u) \; du \; .$$

The functions Λ and S are related by

$$S(t) = e^{-\Lambda(t)} \; .$$

For the sake of simpler notation, assume no ties.
Nelson estimates $\Lambda(t)$ by

$$\hat{\Lambda}(t) = \hat{\Lambda}_2(t) = \sum_{y_{(i)} \leq t} \frac{\delta_{(i)}}{n - i + 1} \; ,$$

and Peterson proposes

$$\hat{\Lambda}_1(t) = \sum_{y_{(i)} \leq t} - \log\left(1 - \frac{\delta_{(i)}}{n - i + 1}\right) \; .$$

The two estimators are very close because for small
x, $\log(1 - x) \cong -x$.   Peterson's estimator corresponds
to the PL estimator of the survival function

$$\hat{S}_1(t) = e^{-\hat{\Lambda}_1(t)} = \prod_{y_{(i)} \leq t} \left(1 - \frac{\delta_{(i)}}{n - i + 1}\right) = \hat{S}(t),$$

while Nelson's estimate corresponds to a different
estimator of the survival function

$$\hat{S}_2(t) = e^{-\hat{\Lambda}_2(t)} \; .$$

Fleming and Harrington recommend $\hat{S}_2(t)$ as an alternative estimator for the survival function and have shown it to have slightly smaller mean square error in some situations.

REFERENCES
Nelson, J. Qual. Tech. (1969).
_____, Technometrics (1972).
Peterson, JASA (1977).
Fleming and Harrington, unpublished manuscript (1979).

Asymptotic Normality. From standard results on (sub)distribution functions,

$$\sqrt{n}[\hat{F}_u(t) - F_u(t)] \overset{W}{\to} Z_{F_u}(t) \ ,$$

$$\sqrt{n}[\hat{H}(t) - H(t)] \overset{W}{\to} Z_H(t) \ ,$$

where Z_{F_u} and Z_H are Gaussian processes.
We expand $\hat{\Lambda}(t)$:

$$\hat{\Lambda}(t) = \int_0^t \frac{d\hat{F}_u(u)}{1 - \hat{H}(u-)} \ ,$$

$$= \int_0^t \left[\frac{1}{1-H} + \frac{\hat{H}-H}{(1-H)^2} + \cdots \right] \left[dF_u + d(\hat{F}_u - F_u) \right] \ ,$$

$$= \int_0^t \frac{dF_u}{1 - H} + \int_0^t \frac{\hat{H} - H}{(1 - H)^2} dF_u$$

$$+ \int_0^t \frac{d(\hat{F}_u - F_u)}{1 - H} + \cdots \ ,$$

$$= \Lambda(t) + \int_0^t \frac{\hat{H} - H}{(1-H)^2} \, dF_u + \frac{(\hat{F}_u - F_u)(t)}{1 - H(t)}$$

$$- \int_0^t \frac{\hat{F}_u - F_u}{(1-H)^2} \, dH + \cdots .$$

The last equality follows from integration by parts of the last integral.   Transposing and multiplying by $\sqrt{n}$,

$$\sqrt{n}[\hat{\Lambda}(t) - \Lambda(t)] = \int_0^t \frac{\sqrt{n}(\hat{H} - H)}{(1-H)^2} \, dF_u$$

$$+ \frac{\sqrt{n}(\hat{F}_u - F_u)(t)}{1 - H(t)}$$

$$- \int_0^t \frac{\sqrt{n}(\hat{F}_u - F_u)}{(1-H)^2} \, dH + \cdots ,$$

$$\xrightarrow{w} \int_0^t \frac{Z_H}{(1-H)^2} \, dF_u + \frac{Z_{F_u}(t)}{1 - H(t)}$$

$$- \int_0^t \frac{Z_{F_u}}{(1-H)^2} \, dH = Z_\Lambda(t) .$$

The limit $Z_\Lambda(t)$, being a weighted average of Gaussian processes, is itself a Gaussian process with

$$E(Z_\Lambda(t)) = 0 ,$$

$$Cov(Z_\Lambda(t_1), Z_\Lambda(t_2)) = \int_0^{t_1 \wedge t_2} \frac{dF_u}{(1-H)^2} .$$

For details, see Breslow and Crowley (1974).
Using this result together with the approximation

$$\hat{S}(t) \cong e^{-\hat{\Lambda}(t)} ,$$

we derive the asymptotic distribution of $\hat{S}(t)$:

$$e^{-\hat{\Lambda}(t)} = e^{-\Lambda(t)} - [\hat{\Lambda}(t) - \Lambda(t)]e^{-\Lambda(t)} + \cdots ,$$

$$\hat{S}(t) \cong S(t) - [\hat{\Lambda}(t) - \Lambda(t)]S(t) + \cdots ,$$

$$\sqrt{n}[\hat{S}(t) - S(t)] \cong -\sqrt{n}[\hat{\Lambda}(t) - \Lambda(t)]S(t) + \cdots ,$$

$$\overset{W}{\to} Z(t) ,$$

where $Z(t)$ is a Gaussian process with

$$E(Z(t)) = 0 ,$$

$$\text{Cov}(Z(t_1), Z(t_2)) = S(t_1) S(t_2) \int_0^{t_1 \wedge t_2} \frac{dF_u}{(1 - H)^2} .$$

REFERENCES
Breslow and Crowley, Ann. Stat. (1974).
Aalen, Scand. J. Stat. (1976).
_____, Ann. Stat. (1978).

4. ROBUST ESTIMATORS

In estimation problems the parameter of interest
can frequently be expressed as a functional

$$\theta = T(F)$$

of the underlying distribution function F.
With no censoring present, the usual estimator is

$$\hat{\theta} = T(F_n) \ ,$$

where F_n is the empirical distribution function.

With censoring present, a reasonable estimator is

$$\hat{\theta} = T(\hat{F}) \ ,$$

where $\hat{F} = 1 - \hat{S}$ and $\hat{S}$ is the PL estimator.

4.1. Mean

$$\theta = T(F) = \int_0^\infty xdF(x) = \int_0^\infty [1 - F(x)]dx$$

$$= \int_0^\infty S(t)dt \ .$$

Without censoring,

$$\hat{\theta} = T(F_n) = \int_0^\infty xdF_n(x) = \bar{x} = \int_0^\infty [1 - F_n(x)]dx \ .$$

With censoring,

$$\hat{\theta} = T(\hat{F}) = \int_0^\infty xd\hat{F}(x) = \int_0^\infty \hat{S}(t)dt \ .$$

$$\text{AVar}(\hat{\theta}) = \frac{1}{n} \int_0^\infty \frac{1}{[1 - H(s)]^2} \left(\int_s^\infty S(u)du \right)^2 dF_u(s) \ .$$

With no ties present,

$$\widehat{AVar}(\hat{\theta}) = \sum_{i=1}^{n} \left(\int_{y_{(i)}}^{\infty} \hat{S}(u)du \right)^2 \frac{\delta_{(i)}}{(n-i)(n-i+1)} \, .$$

Immediately, we have a difficulty. If $y_{(n)}$ is censored, then $\hat{S}(t)$ does not approach zero as $t \to \infty$, so the integrals are infinite.

We discuss three possible solutions.

1. Redefinition of last observation. Change $\delta_{(n)} = 0$ to $\delta_{(n)} = 1$. We illustrate with the maintained AML data of Embury et al.

$$\hat{\theta} = \quad 9 \times .091 + 13 \times .091 + 18 \times .102$$
$$+23 \times .102 + 31 \times .123 + 34 \times .123$$
$$+48 \times .184 + (161 \times .184) \, ,$$
$$= 23.011 + (29.624) \, ,$$
$$= 52.635 \, .$$

The tail, and in particular the last observation, has heavy weight. This is due both to the PL estimator putting increased weights on the last observations and to the skewness of the distribution.

2. Restricted mean (Meier and Sander). For fixed s_0 define a mean over $(0, s_0]$ and estimate it by

$$\hat{\theta} = \int_0^{s_0} \hat{S}(t)dt \, .$$

3. Variable upper limit (Susarla and Van Ryzin). Estimate

$$\theta = \int_0^\infty S(t)dt$$

with

$$\hat{\theta} = \int_0^{s_n} \hat{S}(t)dt \ ,$$

where $\{s_n\}$ is a sequence of numbers converging monotonically to ∞. Unfortunately, the proper choice of s_n depends on F, G and there exist no good guidelines for use in practice as yet.

REFERENCES
Kaplan and Meier, JASA (1958).
Meier, Perspectives in Prob. and Stat. (1975).
Sander, Stanford Univ. Tech. Report No. 8 (1975).
Susarla and Van Ryzin, Ann. Stat. (1980).

4.2. L-Estimators

A basic assumption when using L-estimators is that the underlying distribution F is symmetric about θ. Typically survival times do not have a symmetric distribution because they are positive. However, before estimating we can symmetrize the data by applying a transformation, as, for example, by taking logarithms.
An L-estimator is of the form

$$\hat{\theta} = \int_{-\infty}^\infty xJ(\hat{F}(x))d\hat{F}(x) \ ,$$

where J , defined on [0, 1] , is symmetric about 1/2 and satisfies $\int_0^1 J(u)du = 1$. An important

L-estimator is the trimmed mean with

$$J(u) = \frac{1}{1 - 2\alpha} \, I_{[\alpha, \, 1-\alpha]}(u) \ .$$

With censored data the asymptotic variance of an L-estimator is

$$AVar(\hat{\theta}) = \frac{1}{n} \int_{-\infty}^{\infty} \int_{-\infty}^{\infty} S(t) \, J(S(t)) \, S(u)$$

$$\cdot \, J(S(u)) \left\{ \int_{-\infty}^{t \wedge u} \frac{dF_u(s)}{[1 - H(s)]^2} \right\} dt du \ .$$

REFERENCES
Sander, Stanford Univ. Tech. Report No. 8 (1975).
Reid, <u>Ann. Stat.</u> (1981).

4.3. M-Estimators

Again, a basic assumption is that F is symmetric, so transform the data first. An M-estimator $\hat{\theta}$ is the solution to

$$\int_{-\infty}^{\infty} \psi(x - \hat{\theta}) \, d\hat{F}(x) = 0 \ .$$

The function $\psi(x - \theta)$ generalizes $f'(x - \theta)/f(x - \theta)$ so M-estimators generalize maximum likelihood estimators. The Tukey biweight estimator corresponds to

$$\psi(x) = \begin{cases} x(1 - x^2)^2 & \text{if } |x| \leq 1 \ , \\ 0 & \text{if } |x| > 1 \ . \end{cases}$$

In actual applications the data would need to be scaled by an appropriate scale estimator.

With censored data the asymptotic variance of an M-estimator is

$$AVar(\hat{\theta}) = \frac{1}{n} \int_{-\infty}^{\infty} \frac{1}{[1 - H(s)]^2}$$

$$\cdot \left(\int_{s}^{\infty} \frac{1}{E\psi'} S(t) \ \psi'(t - \theta)dt \right)^2 dF_u(s) \ ,$$

where

$$E\psi' = \int_{-\infty}^{\infty} \psi'(t - \theta) \ dF(t) \ .$$

REFERENCE
Reid, <u>Ann. Stat.</u> (1981).

At this point in time L- and M-estimators with censored data are experimental.   Their virtues and defects have not been established.   However, the median estimator is frequently used.

4.4.   Median

$$\theta = S^{-1}\left(\frac{1}{2}\right) \ .$$

A reasonable estimator for θ is

$$\hat{\theta} = \hat{S}^{-1}\left(\frac{1}{2}\right) \ .$$

If $\hat{S}^{-1}(1/2)$ does not have a unique solution, then define $\hat{\theta}$ to be the midpoint of the interval consisting of the solutions.

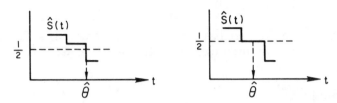

Empirical evidence suggests that this straight-forward estimator tends to be too large. The PL estimator gives increasing jump sizes with increasing t , and due to censored observations dropping out, the gaps between uncensored observations tend to increase with t.   Therefore, $\hat{\theta}$ tends to be too large.

A possible way to alleviate this problem is to define $\hat{\hat{S}}(t)$, a linear smooth of $\hat{S}(t)$, and $\hat{\hat{\theta}} = \hat{\hat{S}}^{-1}(1/2)$.

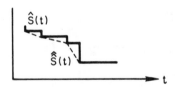

For example, in the maintained AML data of Embury et al.

$$\hat{S}(23) = .614 \, ,$$

$$\hat{S}(31) = .491 \, ,$$

$$\hat{\hat{\theta}} = 31 - \frac{(8)(.009)}{(.123)} = 30.415 \, .$$

We need the variance of $\hat{\theta}$. The asymptotic variance is

$$\text{AVar}(\hat{\theta}) = \frac{\text{AVar}(\hat{S}(\theta))}{f^2(\theta)} \ .$$

$\text{AVar}(\hat{S}(\theta))$ can be estimated using Greenwood's formula, but f is an unknown density and is difficult to estimate.

REFERENCES
Sander, Stanford Univ. Tech. Report No. 5 (1975), discusses the asymptotic variance.
Földes, Rejto, and Winter, unpublished manuscript (1978), discuss density estimation using censored data.
Reid, <u>Ann. Stat.</u> (1981), discusses the asymptotic variance.
_____ and Iyengar, unpublished notes (1979), consider estimates of the variance.
Efron, Stanford Univ. Tech. Report No. 53 (1980), uses the bootstrap to measure the variability of $\hat{\theta}$.

5. BAYES ESTIMATORS

Assume no ties.   Denoting $N_y(t) = \#(y_i > t)$,

$$\hat{S}(t) = \prod_{y_{(i)} \leq t} \left[\frac{n - i}{n - i + 1} \right]^{\delta(i)} \ ,$$

$$= \prod_{y_{(i)} \leq t} \left[\frac{n - i + 1}{n - i} \right]^{-\delta(i)}$$

$$\cdot \frac{1}{n} \left\{ \frac{n}{n - 1} \cdot \frac{n - 1}{n - 2} \cdot \ \cdots \ \frac{N_y(t) + 1}{N_y(t)} \right\}^{N_y(t)}_{1} \ ,$$

$$= \frac{N_y(t)}{n} \prod_{y_{(i)} \le t} \left[\frac{n - i + 1}{n - i} \right]^{1-\delta}(i) \quad .$$

Susarla and Van Ryzin show that the Bayes estimator of $S(t)$ has a similar form:

$$\hat{S}_\alpha(t) = \frac{\alpha(t, \infty) + N_y(t)}{\alpha(0, \infty) + n}$$

$$\times \prod_{y_{(i)} \le t} \left[\frac{\alpha[y_{(i)}, \infty) + (n - i + 1)}{\alpha[y_{(i)}, \infty) + (n - i)} \right]^{1-\delta}(i) \quad .$$

The estimator $\hat{S}_\alpha(t)$ is the Bayes estimator under the loss function

$$L(\hat{\delta}, S) = \int_0^\infty [\hat{\delta}(t) - S(t)]^2 \, dw(t) \quad ,$$

where w is any nonnegative nondecreasing function, and with a Dirichlet process prior $\mathcal{P}_\alpha$ with parameter α on the family $\{P\}$ of all possible distributions. The parameter α is a finite nonnegative measure on $(0, \infty)$.

We say that the random probability measure P has a <u>Dirichlet process prior</u> with parameter α if for any measurable partition $B_1, \ldots, B_k$ of $(0, \infty)$,

$$(P(B_1), \ldots, P(B_k))$$

$$\sim \text{Dirichlet}(\alpha(B_1), \ldots, \alpha(B_k)) \quad .$$

Recall that the Dirichlet$(\alpha_1, \ldots, \alpha_k)$ distribution has density

$$f(x_1, \ldots, x_k) \propto x_1^{\alpha_1 - 1} \, x_2^{\alpha_2 - 1} \cdots x_k^{\alpha_k - 1}$$

$$\cdot \, I(x_i \geq 0, \, x_1 + \cdots + x_k = 1) \; .$$

Notice that for $k = 2$, the Dirichlet distribution is just the beta distribution.

Under a parametric model, we assume our observation X has distribution P_θ where θ is picked by nature according to some prior distribution. In the nonparametric situation, we observe T with distribution P where P is picked by nature according to the distribution $\mathcal{P}_\alpha$. In other words, our survival time T is obtained by $\mathcal{P}_\alpha$ generating P and P generating T. It can be shown that

$$\Pr\{T \in A\} = \frac{\alpha(A)}{\alpha(0, \, \infty)} \; . \tag{10}$$

Equation (10) gives an interpretation to the parameter α. The ratio $\alpha(A)/\alpha(0, \infty)$ is our prior guess on the probability of the set A. For example, if we believe T has exponential distribution with mean $1/\lambda_0$, then

$$\frac{\alpha(t, \, \infty)}{\alpha(0, \, \infty)} = e^{-\lambda_0 t} \; .$$

Also, the total mass $\alpha(0, \infty)$ represents the strength of our prior belief. For example, $\alpha(0, \infty) = 10$ says our prior belief is worth ten observations.

Return to the case where

$$\frac{\alpha(t, \, \infty)}{\alpha(0, \, \infty)} = e^{-\lambda_0 t} \; .$$

Then $\hat{S}_\alpha(t)$ compares with $\hat{S}(t)$ in the following way:

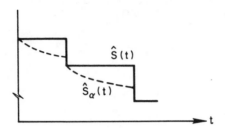

Rai, Susarla, and Van Ryzin show that in many cases, $\hat{S}_\alpha$ gives a smaller mean square error than $\hat{S}$, even when the prior is incorrect.

In case of ties, the Bayes estimate is

$$\hat{S}_\alpha(t) = \frac{\alpha(t, \infty) + N_y(t)}{\alpha(0, \infty) + n}$$

$$\times \prod_{y'_{(j)} \leq t} \left[\frac{\alpha[y'_{(j)}, \infty) + N_y(y'_{(j)}-)}{\alpha[y'_{(j)}, \infty) + N_y(y'_{(j)})}\right]^{1-\delta'(j)} .$$

REFERENCES

Ferguson, Ann. Stat. (1973), discusses the Dirichlet process prior.

Susarla and Van Ryzin, JASA (1976), derive the Bayes estimate in the censored case.

_____ and _____, Ann. Stat. (1978b), study the asymptotic behavior of Bayes estimates.

Ferguson and Phadia, Ann. Stat. (1979), examine more general prior distributions.

Rai, Susarla, and Van Ryzin, Comm. Stat. B (1980), look at mean square errors.

<u>Empirical Bayes Estimators</u>.   Instead of using a prior guess α , we could use the sample to estimate α.

REFERENCES
Susarla and Van Ryzin, <u>Ann. Stat.</u> (1978a).
Phadia, <u>Ann. Stat.</u> (1980).

FOUR

NONPARAMETRIC METHODS: TWO SAMPLES

For the first sample, let T_1, T_2, ..., T_m be iid each with d.f. F_1, and C_1, C_2, ..., C_m be iid each with d.f. G_1. C_i is the censoring time associated with T_i. We can observe (X_1, δ_1), ..., (X_m, δ_m) where

$$X_i = T_i \wedge C_i, \quad \delta_i = I(T_i \le C_i) .$$

For the second sample, let U_1, U_2, ..., U_n be iid each with d.f. F_2, and D_1, D_2, ..., D_n be iid each with d.f. G_2. D_j is the censoring time associated with U_j, and we observe (Y_1, ε_1), ..., (Y_n, ε_n) where

$$Y_j = U_j \wedge D_j, \quad \varepsilon_j = I(U_j \le D_j) .$$

The usual two sample problem is to test

$$H_0: F_1 = F_2 \; .$$

Example. Hypothetical Clinical Trial. In the hypothetical clinical trial constructed by Byron Wm. Brown, Jr. in Figures 4a and 4b, let the treatment A patients be the X observations and the treatment B patients be the Y observations.

Rx A: 3,   5,   7,   9+, 18

Rx B: 12, 19, 20, 20+, 33+ .

1.   GEHAN TEST

Gehan's test is an extension of the Wilcoxon test. Let the observations from the two samples be

$$X_1, \; \ldots, \; X_m; \; Y_1, \; \ldots, \; Y_n \; .$$

Order the combined sample and define

$$Z_{(1)} < Z_{(2)} < \cdots < Z_{(m+n)}$$

$$R_{1i} = \text{rank of } X_i \; ,$$

$$R_1 = \sum_{i=1}^{m} R_{1i} \; .$$

Reject H_0 if R_1 is too small or too large. Use small sample tables or the large sample approximation

$$\frac{R_1 - E_0(R_1)}{\sqrt{\text{Var}_0(R_1)}} = \frac{R_1 - \dfrac{m(m + n + 1)}{2}}{\sqrt{\dfrac{mn(m + n + 1)}{12}}} \overset{a}{\sim} N(0, 1) \; ,$$

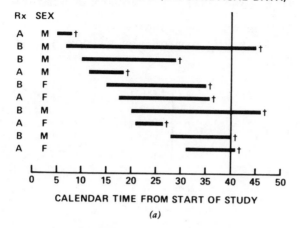

Figure 4a. Survival times for 10 cancer patients randomly assigned to treatments A and B (hypothetical data).

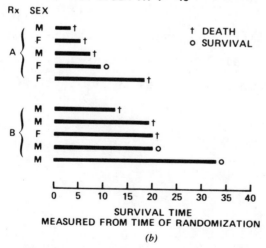

Figure 4b. Survival times from time of randomization for 10 cancer patients – assuming termination of study at t = 40.

where $E_0(R_1)$ and $Var_0(R_1)$ are the moments calculated under the null hypothesis.

The <u>Mann-Whitney form</u> of the Wilcoxon test will be
useful.   Define

$$U(X_i, Y_j) = U_{ij} = \begin{cases} +1 & \text{if } X_i > Y_j , \\ 0 & \text{if } X_i = Y_j , \\ -1 & \text{if } X_i < Y_j , \end{cases}$$

$$U = \sum_{i=1}^{m} \sum_{j=1}^{n} U_{ij} .$$

It can be shown that

$$R_1 = \frac{m(m + n + 1)}{2} + \frac{1}{2} U .$$

To see this notice that if we have the total separa-
tion $x_{(1)} < \cdots < x_{(m)} < y_{(1)} < \cdots < y_{(n)}$, then
$R_1 = m(m + 1)/2$. For every interchange of a conti-
guous x, y pair, R_1 is increased by 1, and the
number of such interchanges is $\Sigma_i \Sigma_j (1/2)(U_{ij} + 1)$.
Therefore,

$$R_1 = \frac{m(m + 1)}{2} + \sum_i \sum_j \frac{1}{2}(U_{ij} + 1) ,$$

$$= \frac{m(m + 1)}{2} + \frac{mn}{2} + \frac{1}{2} U ,$$

$$= \frac{m(m + n + 1)}{2} + \frac{1}{2} U .$$

The Mann-Whitney test rejects H_0 if U or $|U|$
is too large. Use small sample tables or the large

sample approximation

$$\frac{U - E_0(U)}{\sqrt{Var_0(U)}} = \frac{U}{\sqrt{\dfrac{mn(m+n+1)}{3}}} \overset{a}{\sim} N(0, 1) .$$

For censored data, Gehan defines

$$U_{ij} = \begin{cases} 1 & \text{if we know } t_i > u_j \text{ , that is,} \\ & (x_i > y_j, \; \varepsilon_j = 1) \text{ or} \\ & (x_i = y_j, \; \delta_i = 0, \; \varepsilon_j = 1) \text{ ,} \\ 0 & \text{otherwise ,} \\ -1 & \text{if we know } t_i < u_j \text{ , that is,} \\ & (x_i < y_j, \; \delta_i = 1) \text{ or} \\ & (x_i = y_j, \; \delta_i = 1, \; \varepsilon_j = 0) \text{ ,} \end{cases}$$

$$U = \sum_{i=1}^{m} \sum_{j=1}^{n} U_{ij} .$$

Reject H_0 if U or $|U|$ is large. The statistic U is asymptotically normally distributed by the theory of two-sample U-statistics, but to calculate the critical values we need to know the moments of U.

1.1. Mean and Variance of U

With no censoring, the mean and variance can be calculated using permutation theory. Under H_0 , consider sampling m balls without replacement from an urn containing $m + n$ balls labeled Z_1, ..., Z_{m+n}. Think of the labels on the m sampled balls as the values of X_1, ..., X_m , and the labels on the

n unsampled balls as the values of Y_1, ..., Y_n.
Let $E_{0,P}(U)$ and $Var_{0,P}(U)$ be the moments under
this permutation model.   Then,

$$E_{0,P}(U) = 0 = E_0(U) \; ,$$

$$Var_{0,P}(U) = \frac{mn(m + n + 1)}{3} = Var_0(U) \; .$$

With censoring, Gehan also uses permutation theory
but under the more restrictive null hypothesis

$$H_0^* : F_1 = F_2 \quad \text{and} \quad G_1 = G_2 \; .$$

Let the combined sample be denoted by

$$(Z_1, \zeta_1), \; \ldots, \; (Z_{m+n}, \zeta_{m+n}) \; .$$

Consider sampling m balls without replacement from
an urn containing $m + n$ balls labeled (Z_1, ζ_1),
..., (Z_{m+n}, ζ_{m+n}).   Think of the labels on the m
sampled balls as (X_1, δ_1), ..., (X_m, δ_m) and the
labels on the n unsampled balls as (Y_1, ε_1), ...,
(Y_n, ε_n).   Then,

$$E_{0,P}^*(U) = 0$$

$$Var_{0,P}^*(U) = (4.3) \quad \text{on page 206 of Gehan (1965)}.$$

The latter is a complicated expression which we will
not record here because Mantel's computational form
for $Var_{0,P}^*(U)$ is easier to work with.

1.2. Mantel Computational Form for $\text{Var}^*_{0,P}(U)$

$$U_{k\ell} = U((Z_k, \zeta_k), (Z_\ell, \zeta_\ell))$$

$$= \begin{cases} +1 & \text{if } (Z_k > Z_\ell,\ \zeta_\ell = 1) \text{ or} \\ & (Z_k = Z_\ell,\ \zeta_k = 0,\ \zeta_\ell = 1)\ , \\ 0 & \text{otherwise}\ , \\ -1 & \text{if } (Z_k < Z_\ell,\ \zeta_k = 1) \text{ or} \\ & (Z_k = Z_\ell,\ \zeta_k = 1,\ \zeta_\ell = 0)\ , \end{cases}$$

$$U^*_k = \sum_{\substack{\ell=1 \\ \neq k}}^{m+n} U_{k\ell}\ ,$$

$$U = \sum_{k=1}^{m+n} U^*_k\ I(k \in I_1)\ ,$$

where I_1 is the set of integers comprising sample 1. Notice that U is equal to Gehan's statistic because $U_{k_1 k_2} = -U_{k_2 k_1}$ so if $k_1, k_2 \in I_1$, they cancel each other out in the sum.

To calculate the permutation distribution of U, suppose we are given $U^*_1, \ldots, U^*_{m+n}$. Under H^*_0, we sample m of these U^*_k without replacement and form U, the sum of these m values. Using results on sampling from finite populations,

$$\text{Var}^*_{0,P}(U) = m\left(\frac{1}{m+n-1}\sum_{i=1}^{m+n}(U^*_i)^2\right)\left(1 - \frac{m}{m+n}\right)\ ,$$

$$= \frac{mn}{(m + n)(m + n - 1)} \sum_{i=1}^{m+n} (U_i^*)^2 \; .$$

1.3. Example

For Brown's hypothetical clinical trial (Figures 4a and 4b):

Z	Rx	# < Z	# > Z	U^*
3	A	0	9	-9
5	A	1	8	-7
7	A	2	7	-5
9+	A	3	0	3
12	B	3	5	-2
18	A	4	4	0
19	B	5	3	+2
20	B	6	2	+4
20+	B	7	0	+7
33+	B	7	0	+7

$$U = -9 - 7 - 5 + 3 + 0 = -18 \; ,$$

$$E_{0,P}^*(U) = 0 \; ,$$

$$Var_{0,P}^*(U) = \frac{(5)(5)(286)}{(10)(9)} = 79.44 \; .$$

Under H_0^* ,

$$\frac{U}{\sqrt{\mathrm{Var}^*_{0,P}(U)}} = \frac{-18}{8.91} = -2.02 \overset{a}{\sim} N(0, 1) \ ,$$

so $P = .022$ is the one-tailed P-value.

REFERENCES
Gehan, Biometrika (1965).
Mantel, Biometrics (1967).

1.4.   Variance under H_0

The above results on the variance were derived under the assumption $H_0^* : F_1 = F_2$, $G_1 = G_2$. What is
the permutation variance under $H_0 : F_1 = F_2$ with the
censoring patterns held fixed? Suppose

$$V_1, \ \ldots, \ V_{m+n}$$

is the combined sample of $T_1, \ \ldots, \ T_m, \ U_1, \ \ldots, \ U_n$.
Under H_0 we sample from the V_k without replace-
ment and put them in the slots

$$(\underline{\ \ }, \ C_1), \ \ldots, \ (\underline{\ \ }, \ C_m); \ (\underline{\ \ }, \ D_1), \ \ldots, \ (\underline{\ \ }, \ D_n) \ .$$

From here, we want to form

$$(X_1, \ \delta_1), \ \ldots, \ (X_m, \ \delta_m); \ (Y_1, \ \varepsilon_1), \ \ldots, \ (Y_n, \ \varepsilon_n),$$

but unfortunately, because not all the T_i, C_i, U_j,
D_j are observable, not all the $(X_i, \ \delta_i)$, $(Y_j, \ \varepsilon_j)$
can be constructed.

Hyde compares $E_0(\mathrm{Var}^*_{0,P}(U))$ with $\mathrm{Var}_0(U)$.

$$\text{Var}_0(U) = E_0(U^2) \; ,$$

$$= E\left\{\left(\sum_{i=1}^{m} \sum_{j=1}^{n} U_{ij}\right)^2\right\} \; ,$$

$$= mn \; E_0(U_{ij}^2) + mn(n-1) \; E_0(U_{ij}U_{ij'})$$

$$+ \; m(m-1)n \; E_0(U_{ij}U_{i'j})$$

$$+ \; m(m-1)n(n-1) \; E_0(U_{ij}U_{i'j'}) \; .$$

$E_0(\text{Var}_{0,P}^{*}(U)) = $ sum of similar terms.

Letting $m, n \to \infty$ such that $m/(m+n) \to \lambda$, where $0 < \lambda < 1$,

$$R^2 = \underset{m,n\to\infty}{\text{p-lim}} \; \frac{\text{Var}_{0,P}^{*}(U)}{\text{Var}_0(U)} \; ,$$

$$= \lim_{m,n\to\infty} \frac{E_0(\text{Var}_{0,P}^{*}(U))}{\text{Var}_0(U)} \; ,$$

$$= 3\lambda(1-\lambda)$$

$$+\left\{\lambda^3 \; P\{C_1 \wedge C_2 \wedge C_3 > T_1 \wedge T_2 \wedge T_3\}\right.$$

$$+ \; (1-\lambda)^3 \; P\{D_1 \wedge D_2 \wedge D_3 > T_1 \wedge T_2 \wedge T_3\}\Big\}$$

$$\times\left\{\lambda \; P\{C_1 \wedge C_2 \wedge D_1 > T_1 \wedge T_2 \wedge T_3\}\right.$$

$$+ \; (1-\lambda) \; P\{C_1 \wedge D_1 \wedge D_2 > T_1 \wedge T_2 \wedge T_3\}\Big\}^{-1} \; .$$

(11)

From (11) we see that $R^2 > 3\lambda(1-\lambda)$. If $\lambda = 1/2$, then $R^2 > 3/4$, so $R > .87$. Thus, if the sample sizes are equal, $SD^*_{0,P}(U)$ cannot be very much smaller than $SD_0(U)$, no matter what the censoring patterns are.

Suppose the censoring distributions are Lehmann alternatives, that is,

$$(1 - G_1)^{r_1} = 1 - F ,$$

$$(1 - G_2)^{r_2} = 1 - F ,$$

where r_1 and r_2 are related to

$$P_1 = P\{C_1 < T_1\} = P\{\text{Observation being censored in Population 1}\} ,$$

$$P_2 = P\{D_1 < U_1\} = P\{\text{Observation being censored in Population 2}\} ,$$

through

$$P_1 = \frac{1}{r_1 + 1} , \quad P_2 = \frac{1}{r_2 + 1} .$$

In Table 2, Hyde reports R for $\lambda = .5$ under Lehmann alternatives for varying levels of censoring probabilities P_1 and P_2. The table has been partitioned to identify cases in which $|R-1| < .05$. Table 3 is analogous to Table 2 with $\lambda = .2$.

We see from Table 2 that for equal sample sizes, the Gehan test (which assumes equal censoring patterns) uses an approximately correct standard deviation even when the censoring probabilities differ appreciably. Moreover, even when one sample size is four times the other (Table 3), Gehan's test is using

TABLE 2. VALUES OF R WHEN CENSORING DISTRIBUTIONS ARE LEHMANN ALTERNATIVES AND λ = .5

P_2	\ P_1	0.00	0.10	0.20	0.30	0.40	0.50	0.60	0.70	0.80	0.90
0.00		1.00	1.00	1.00	1.00	1.01	1.01	1.02	1.04	1.09	1.20
0.10		1.00	1.00	1.00	1.00	1.00	1.01	1.02	1.04	1.07	1.18
0.20		1.00	1.00	1.00	1.00	1.00	1.01	1.01	1.03	1.06	1.16
0.30		1.00	1.00	1.00	1.00	1.00	1.00	1.01	1.02	1.05	1.13
0.40		1.01	1.00	1.00	1.00	1.00	1.00	1.00	1.01	1.04	1.11
0.50		1.01	1.01	1.01	1.00	1.00	1.00	1.00	1.01	1.02	1.09
0.60		1.02	1.02	1.01	1.01	1.00	1.00	1.00	1.00	1.01	1.06
0.70		1.04	1.04	1.03	1.02	1.01	1.01	1.00	1.00	1.00	1.04
0.80		1.09	1.07	1.06	1.05	1.04	1.02	1.01	1.00	1.00	1.01
0.90		1.20	1.18	1.16	1.13	1.11	1.09	1.06	1.04	1.01	1.00

TABLE 3. VALUES OF R WHEN CENSORING DISTRIBUTIONS ARE LEHMANN ALTERNATIVES AND λ = .2

p_2	p_1									
	0.00	0.10	0.20	0.30	0.40	0.50	0.60	0.70	0.80	0.90
0.00	1.00	1.01	1.02	1.04	1.06	1.09	1.14	1.21	1.33	1.65
0.10	0.99	1.00	1.01	1.03	1.05	1.08	1.12	1.18	1.29	1.59
0.20	0.98	0.99	1.00	1.01	1.03	1.06	1.09	1.15	1.26	1.53
0.30	0.97	0.98	0.99	1.00	1.02	1.04	1.07	1.12	1.22	1.47
0.40	0.96	0.97	0.97	0.99	1.00	1.02	1.05	1.09	1.18	1.40
0.50	0.95	0.95	0.96	0.97	0.98	1.00	1.02	1.06	1.14	1.33
0.60	0.94	0.94	0.95	0.96	0.97	0.98	1.00	1.03	1.09	1.26
0.70	0.93	0.93	0.93	0.94	0.95	0.96	0.97	1.00	1.05	1.18
0.80	0.92	0.92	0.92	0.93	0.93	0.94	0.95	0.97	1.00	1.09
0.90	0.92	0.92	0.92	0.92	0.92	0.92	0.92	0.93	0.95	1.00

a nearly correct standard deviation for a wide range
of censoring probabilities.

REFERENCES

Gilbert, Univ. Chicago thesis (1962), was the
   first to calculate $Var_0(U)$.

Hyde, Stanford Univ. Tech. Report No. 30 (1977).

2. MANTEL-HAENSZEL TEST

2.1. Single 2×2 Table

Suppose we have two populations, and an individual
in either population can have one of two characteris-
tics. For example, Population 1 might be cancer patients under a certain treatment and Population 2
cancer patients under a different treatment. The
patients in either group may either die within a year
or survive beyond a year. The data may be summarized
in a 2×2 table.

	Dead	Alive	
Population 1	a	b	n_1
Population 2	c	d	n_2
	m_1	m_2	n

Denote

$$p_1 = P\{Dead | Population\ 1\}\ ,$$

$$p_2 = P\{Dead | Population\ 2\}\ .$$

To test

$$H_0 : p_1 = p_2\ ,$$

use the statistic

$$
\chi^2 = \left[\frac{\hat{p}_1 - \hat{p}_2}{\sqrt{\hat{p}(1 - \hat{p})(1/n_1 + 1/n_2)}}\right]^2 ,
$$

$$
= \frac{n(ad - bc)^2}{n_1 n_2 m_1 m_2} ,
$$

where

$$
\hat{p}_1 = \frac{a}{n_1} , \quad \hat{p}_2 = \frac{c}{n_2} , \quad \hat{p} = \frac{m_1}{n} ,
$$

or, including the continuity correction,

$$
\chi_c^2 = \frac{n(|ad - bc| - n/2)^2}{n_1 n_2 m_1 m_2} .
$$

χ^2 is approximately distributed as χ_1^2. This is an approximation to the exact discrete conditional distribution under H_0. Given n_1, n_2, m_1, m_2 fixed, the random variable A, which is the entry in the 1, 1 cell of the 2×2 table, has a hypergeometric distribution

$$
P\{A = a\} = \frac{\binom{n_1}{a}\binom{n_2}{m_1 - a}}{\binom{n}{m_1}} .
$$

The first two moments of the hypergeometric distribution are

$$E_0(A) = \frac{n_1 m_1}{n} ,$$

$$Var_0(A) = \frac{n_1 n_2 m_1 m_2}{n^2(n-1)} .$$

Consequently,

$$ad - bc = n(a - E_0(A)) ,$$

$$n_1 n_2 m_1 m_2 = n^2(n - 1) \, Var_0(A) ,$$

$$\chi^2 = \frac{n(ad - bc)^2}{n_1 n_2 m_1 m_2} = \frac{n}{n-1} \left[\frac{a - E_0(A)}{\sqrt{Var_0(A)}} \right]^2 .$$

2.2. Sequence of 2 × 2 Tables

Now suppose we have a sequence of 2 × 2 tables. For example, we might have k hospitals; at each hospital, patients receive either Treatment 1 or Treatment 2 and their responses are observed.

Because there may be differences among hospitals, we do not want to combine all k tables into a single 2 × 2 table. We want to test

$$H_0 : \quad P_{11} = P_{12}, \ \cdots, \ P_{k1} = P_{k2} ,$$

where

$$P_{i1} = P\{Dead \,|\, Treatment\ 1,\ Hospital\ i\} ,$$

$$P_{i2} = P\{Dead \,|\, Treatment\ 2,\ Hospital\ i\} .$$

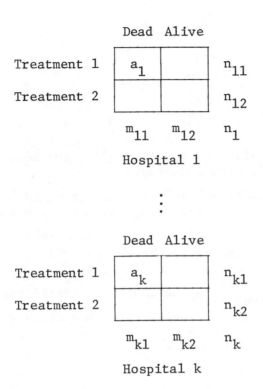

Hospital 1

Hospital k

Use the <u>Mantel-Haenszel statistic</u>

$$MH = \frac{\sum\limits_{i=1}^{k} (a_i - E_0(A_i))}{\sqrt{\sum\limits_{i=1}^{k} Var_0(A_i)}} \ .$$

Including the continuity correction, the Mantel-Haenszel statistic is

$$
MH_c = \frac{\left| \sum_{i=1}^{k} (a_i - E_0(A_i)) \right| - \frac{1}{2}}{\sqrt{\sum_{i=1}^{k} Var_0(A_i)}} .
$$

Lininger et al. show that including the continuity correction is very conservative.

If the tables are independent, then $MH \overset{a}{\sim} N(0, 1)$ either when k is fixed and $n_i \to \infty$ or when $k \to \infty$ and the tables are also identically distributed.

In survival analysis the MH statistic is applied as follows. Recall that $(Z_{(1)}, \zeta_{(1)}), \ldots,$ $(Z_{(m+n)}, \zeta_{(m+n)})$ is the combined ordered sample. Construct a 2×2 table for each uncensored time point. Compute the MH statistic for this sequence of tables to test $H_0 : F_1 = F_2$.

These tables are not independent because, for example, $\Re(z_{(1)})$ and $\Re(z_{(3)})$ almost coincide. But asymptotic normality still holds, as we will argue below. Varying censoring patterns have no effect on the MH statistic.

This test is also called the <u>log rank test</u> (see Chapter 6, Section 2.1).

2.3.   Example

The computations for the MH statistic in Brown's hypothetical clinical trial (Figures 4a and 4b) are given in Table 4. The column labeled z contains the uncensored ordered observations. The next four columns labeled $n, m_1, n_1,$ a construct the 2×2 tables. The next column is $E_0(A) = n_1 m_1 / n$. The product of the last two columns, labeled

The Mantel-Haenszel sequence of 2×2 tables is illustrated below:

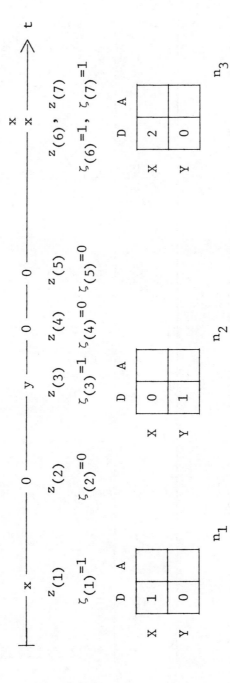

TABLE 4. COMPUTATIONS FOR THE MANTEL-HAENSZEL STATISTIC IN BROWN'S HYPOTHETICAL CLINICAL TRIAL

z	n	m_1	n_1	a	$E_0(A)$	$a - E_0(A)$	$\dfrac{m_1(n - m_1)}{n - 1}$	$\dfrac{n_1}{n}\left(1 - \dfrac{n_1}{n}\right)$
3	10	1	5	1	.50	.50	1	.2500
5	9	1	4	1	.44	.56	1	.2469
7	8	1	3	1	.38	.62	1	.2344
12	6	1	1	0	.17	-.17	1	.1389
18	5	1	1	1	.20	.80	1	.1600
19	4	1	0	0	0	0	1	0
20	3	1	0	0	0	0	1	0

$m_1(n - m_1)/(n - 1)$ and $(n_1/n)(1 - n_1/n)$, is $\mathrm{Var}_0(A)$; it is convenient to break up the calculation of $\mathrm{Var}_0(A)$ in this way because $m_1(n - m_1)/(n-1)$ is usually equal to 1 and $(n_1/n)(1 - n_1/n)$ is the product of the proportions in the two samples.

$$MH = \frac{\text{sum of } a - E_0(A) \text{ column}}{\sqrt{\text{sum of } \left(\frac{m_1(n-m_1)}{n-1} \text{ col.} \times \frac{n_1}{n}\left(1 - \frac{n_1}{n}\right) \text{ col.}\right)}} ,$$

$$= \frac{2.31}{1.02} = 2.26 .$$

$$P = .012 \text{ (one-tailed)} .$$

$$MH_c = \frac{2.31 - 0.50}{1.02} ,$$

$$= \frac{1.81}{1.02} = 1.77 .$$

$$P = .038 \text{ (one-tailed)} .$$

2.4.   Asymptotic Normality

To show asymptotic normality, we adapt Crowley's representation to our case.   Assume no ties.
    Denote

$$N = m + n ,$$

$$\hat{H}(t) = \frac{1}{N} \sum_{i=1}^{N} I(Z_i \leq t) ,$$

$$\hat{H}_1(t) = \frac{1}{m} \sum_{i=1}^{m} I(X_i \leq t) ,$$

$$\hat{H}_u(t) = \frac{1}{N} \sum_{i=1}^{N} I(Z_i \leq t, \zeta_i = 1) \ ,$$

$$\hat{H}_{1u}(t) = \frac{1}{m} \sum_{i=1}^{m} I(X_i \leq t, \delta_i = 1) \ .$$

Then the numerator of MH is

$$\sum_{i=1}^{k} (a_i - E_0(A_i)) = m \left\{ \int_0^\infty d\hat{H}_{1u}(s) \right.$$

$$\left. - \int_0^\infty \frac{1 - \hat{H}_1(s-)}{1 - \hat{H}(s-)} \ d\hat{H}_u(s) \right\} \ .$$

To see this, recall that $E(A_i) = m_{i1} n_{i1} / n_i$ where a_i, m_{i1}, n_{i1}, n_i are gotten from the 2×2 table corresponding to the i-th uncensored observation:

Because we have assumed no ties, $m_{i1} = 1$. Letting s_i denote the time of the i-th uncensored observation,

$$n_{i1} = \#(X\text{'s remaining at time } s_i-) \ ,$$
$$= m(1 - \hat{H}_1(s_i-)) \ ,$$

$$n_i = \#(Z\text{'s remaining at time } s_i-) ,$$

$$= N(1 - \hat{H}(s_i-)) .$$

Now that we have the numerator of MH expressed in terms of empirical (sub)distribution functions, we may apply arguments similar to those used in showing the asymptotic normality of the PL estimator.

REFERENCES
Mantel and Haenszel, J. Natl. Cancer Inst. (1959).
Crowley, JASA (1974).
Lininger et al., Biometrika (1979).

3. TARONE-WARE CLASS OF TESTS

After constructing a 2×2 table for each uncensored observation, Tarone and Ware suggest weighting each table, forming

$$\sum_{i=1}^{k} w_i[a_i - E_0(A_i)] = \sum_{i=1}^{k} w_i\left[a_i - \frac{m_{i1} n_{i1}}{n_i}\right]. \quad (12)$$

For the variance, use

$$\sum_{i=1}^{k} w_i^2 \, \text{Var}_0(A_i) = \sum_{i=1}^{k} w_i^2 \left[\frac{m_{i1}(n_i - m_{i1})}{n_i - 1}\right]$$

$$\times \left[\left(\frac{n_{i1}}{n_i}\right)\left(1 - \frac{n_{i1}}{n_i}\right)\right]. \quad (13)$$

There are three important special cases:

(i) $w_i \equiv 1$ gives the MH statistic.

(ii) $w_i = n_i$ gives the Gehan statistic.

(iii) $w_i = \sqrt{n_i}$ is suggested by Tarone and Ware.

NOTES

(i)   Which test should we use?   The Gehan statistic puts more weight on the beginning observations, while the MH statistic puts equal weight on each observation.   Tarone and Ware's suggestion is intermediate between the two, and they claim that the weights $w_i = \sqrt{n_i}$ have high efficiency over a range of alternatives.

(ii)   Although (12) is identical to the Gehan statistic U , $\hat{Var}_{TW}(U)$, given by (13), is not the same as $Var^*_{0,P}(U)$.   Asymptotically, $\hat{Var}_{TW}(U)$ is equivalent to the variance of U under H_0 while $Var^*_{0,P}(U)$ is the variance under H_0^*.

Example.   Refer to Table 4 where we calculated the MH statistic for Brown's clinical trial.

$$\sum_{i=1}^{k} n_i(a_i - E_0(A_i)) = (10)(.50) + (9)(.56)$$
$$+ (8)(.62) + (6)(-.17)$$
$$+ (5)(.80) ,$$
$$= 17.98 ,$$

which is what we got for Gehan's statistic U except for sign and roundoff.   Also,

$$\hat{Var}_{TW}(U) = \sum n_i^2 \left[\frac{m_{i1}(n_i - m_{i1})}{n_i - 1} \right] \left[\left(\frac{n_{i1}}{n_i} \right) \left(1 - \frac{n_{i1}}{n_i} \right) \right] ,$$

$$= (10^2)(.25) + (9^2)(.2469)$$

$$+ (8^2)(.2344) + (6^2)(.1389)$$

$$+ (5^2)(.16) \, ,$$

$$= 69 \, ,$$

$$\text{Var}^*_{0,P}(U) = 79.44 \, ,$$

which give

$$\sqrt{\hat{\text{Var}}_{TW}(U)} = 8.31 \quad \text{and} \quad \sqrt{\text{Var}^*_{0,P}(U)} = 8.91 \, .$$

REFERENCE
Tarone and Ware, <u>Biometrika</u> (1977).

4. EFRON TEST

Recall that in the construction of Gehan's test, we defined the score function

$$U_{ij} = \begin{cases} 1 & \text{if we know } t_i > u_j \, , \\ 0 & \text{otherwise} \, , \\ -1 & \text{if we know } t_i < u_j \, . \end{cases}$$

Suppose we have the following situation:

$$\begin{array}{ll} y_j & x_i \\ \varepsilon_j = 0 & \delta_i = 1 \end{array}$$

The Gehan test assigns the score $U_{ij} = 0$ for this pair regardless of how much larger x_i is compared to y_j. Efron suggests using the score

$$U_{ij} = \hat{P}\{T_i > U_j \mid (x_i, \delta_i), (y_j, \epsilon_j)\} .$$

For the depicted situation,

$$U_{ij} = \hat{P}\{U_j < x_i \mid U_j > y_j\} ,$$

$$= \frac{\hat{F}_2(x_i) - \hat{F}_2(y_j)}{1 - \hat{F}_2(y_j)} ,$$

where $\hat{F}_2$ is the Kaplan-Meier PL estimator for Population 2.

Use of these scores along with 1 and 0 instead of 1 and -1 leads to the statistic

$$\int_0^\infty [1 - \hat{F}_1(u)]d\hat{F}_2(u) = \hat{P}\{T_i > U_j\} . \tag{14}$$

The estimator $\hat{P}\{T_i > U_j\}$ is the GMLE of $P\{T_i > U_j\}$, which is the parameter the Wilcoxon statistic is estimating in the uncensored case, that is,

$$\frac{1}{mn} U \overset{a.s.}{\to} P\{X > Y\}$$

in the uncensored case with $U_{ij} = 0$ or 1.

The estimator (14) is somewhat unstable in the tails, which has prevented its widespread use.

REFERENCE
Efron, Proc. Fifth Berkeley Symp. IV (1967).

FIVE

NONPARAMETRIC METHODS: K SAMPLES

For the i-th sample $(i = 1, \ldots, K)$, let $T_{i1}, \ldots,$ T_{in_i} be iid each with d.f. F_i, and $C_{i1}, \ldots, C_{in_i}$ be iid each with d.f. G_i. C_{ij} is the censoring time associated with T_{ij}. We can observe $(X_{i1}, \delta_{i1}), \ldots, (X_{in_i}, \delta_{in_i})$ where

$$X_{ij} = T_{ij} \wedge C_{ij} , \quad \delta_{ij} = I(T_{ij} \leq C_{ij}) .$$

We are interested in the hypothesis

$$H_0 : F_1 = \cdots = F_K .$$

1. GENERALIZED GEHAN TEST (BRESLOW)

Using the score function that we defined for the two sample problem, let

$$W_i = \sum_{j=1}^{n_i} \sum_{\substack{i'=1 \\ \neq i}}^{K} \sum_{j'=1}^{n_{i'}} U((X_{ij}, \delta_{ij}), (X_{i'j'}, \delta_{i'j'})) ,$$

$$\underset{\sim}{W} = (W_1, \ldots, W_K)' .$$

Breslow obtains the asymptotic covariance matrix of $\underset{\sim}{W}$ under the more restrictive hypothesis

$$H_0^* : F_1 = \cdots = F_K ; \quad G_1 = \cdots = G_K .$$

Denote

$$N = \sum_{i=1}^{K} n_i .$$

Assuming $n_i/N \to \lambda_i$ as $N \to \infty$, $i = 1, \ldots, K$,

$$\underset{\sim}{W} \overset{a}{\sim} N(\underset{\sim}{\mu}_0^*, N^3 \underset{\sim}{\Sigma}_0^*) ,$$

where

$$\underset{\sim}{\mu}_0^* = 0 ,$$

$$\underset{\sim}{\Sigma}_0^* = \left(\int_0^\infty [1 - H(u)]^2 \, dH_u(u) \right)$$

$$\times \begin{pmatrix} \lambda_1(1-\lambda_1) & & & \\ & \ddots & & -\lambda_i\lambda_j \\ & & \ddots & \\ -\lambda_i\lambda_j & & & \ddots \\ & & & & \lambda_K(1-\lambda_K) \end{pmatrix} ,$$

and

$$H_i(t) = P\{X_{i1} \leq t\} \, ,$$

$$H_{iu}(t) = P\{X_{i1} \leq t, \ \delta_{i1} = 1\} \, ,$$

$$H(t) = \lambda_1 \, H_1(t) + \cdots + \lambda_K \, H_K(t) \, ,$$

$$H_u(t) = \lambda_1 \, H_{1u}(t) + \cdots + \lambda_K \, H_{Ku}(t) \, .$$

Since the asymptotic covariance matrix depends on unknown parameters, substitute

$$\hat{\lambda}_i = \frac{n_i}{N} \, ,$$

$$\hat{H}(t) = \frac{1}{N} \sum_{i=1}^{K} \sum_{j=1}^{n_i} I(X_{ij} \leq t) \, ,$$

$$\hat{H}_u(t) = \frac{1}{N} \sum_{i=1}^{K} \sum_{j=1}^{n_i} I(X_{ij} \leq t, \ \delta_{ij} = 1) \, .$$

REFERENCE
Breslow, <u>Biometrika</u> (1970).

<u>Types of Tests.</u>

1. <u>Omnibus χ^2 Test.</u> Use

$$\frac{1}{N^3 \displaystyle\int_0^{\infty} (1 - \hat{H})^2 \, d\hat{H}_u} \sum_{i=1}^{K} \frac{W_i^2}{\hat{\lambda}_i} \overset{a}{\sim} \chi_{K-1}^2 \, .$$

This statistic is equivalent to $\underset{\sim}{W}'(\underset{\sim}{\Sigma_0^*})^{-} \underset{\sim}{W}$ where

$(\Sigma_0^*)^-$ is the generalized inverse of Σ_0^*.

   With no censoring, this statistic is asymptotically equivalent to the Kruskal-Wallis statistic. Recall that if R_{ij} is the rank of X_{ij} among all N
observations, and

$$R_i = \sum_{j=1}^{n_i} R_{ij} \ ,$$

$$\bar{R}_i = \frac{1}{n_i} R_i \ ,$$

$$\bar{R}_{\cdot} = \frac{1}{N} \sum_{i=1}^{K} R_i \ ,$$

then the Kruskal-Wallis statistic is

$$\frac{12}{N(N+1)} \sum_{i=1}^{K} n_i (\bar{R}_i - \bar{R}_{\cdot})^2 = \left(\frac{12}{N(N+1)} \sum_{i=1}^{K} \frac{R_i^2}{n_i} \right)$$
$$- 3N(N+1) \ ,$$

which has asymptotic distribution χ^2_{K-1} under H_0.

   2. <u>Test for Trend</u>. Suppose we know that if the
populations are not all equal, then they are <u>ordered</u>.
For example, the populations may correspond to increasing doses of a drug

$$d_1 < \cdots < d_K \ ,$$

where d_i is the dose given to members of Population
i. In other situations it may be known <u>a priori</u> that

the populations should change monotonically if they differ, but there is no numerical covariate.

When a quantitative measure like dose is available, define

$$\underset{\sim}{\ell} = (d_1, \ldots, d_K)' .$$

If a quantitative variable is not available, then define

$$\underset{\sim}{\ell} = (-(K-1), \ldots, -3, -1, +1, +3, \ldots, +(K-1))'$$

if K is even,

$$\underset{\sim}{\ell} = \left(- \frac{(K-1)}{2}, \ldots, -1, 0, +1, \ldots, + \frac{(K-1)}{2}\right)'$$

if K is odd.

Abelson and Tukey suggest the linear-2 or the linear-2-4 contrasts which we illustrate respectively for K even:

$$\underset{\sim}{\ell} = (-2(K-1), -(K-3), -(K-5), \ldots, +(K-5),$$
$$+(K-3), +2(K-1))' ,$$

$$\underset{\sim}{\ell} = (-4(K-1), -2(K-3), -(K-5), \ldots, +(K-5),$$
$$+2(K-3), +4(K-1))' .$$

Renormalize $\underset{\sim}{W}$ by defining

$$\bar{W}_i = \frac{W_i}{n_i(N-n_i)} ,$$

$$\underset{\sim}{\bar{W}} = (\bar{W}_1, \ldots, \bar{W}_K)' ,$$

and let $\underset{\sim}{c}$ be such that

$$\underset{\sim}{\ell}' \underset{\sim}{\bar{W}} = \underset{\sim}{c}' \underset{\sim}{W} .$$

Then,

$$\frac{\underset{\sim}{c}' \underset{\sim}{W}}{\sqrt{N^3 \underset{\sim}{c}' \underset{\sim}{\Sigma}_0^* \underset{\sim}{c}}} \overset{a}{\sim} N(0, 1) \ ,$$

and this statistic can be used to test

$$H_0 : \ F_1 = \cdots = F_K \quad \text{against} \quad H_1 : \ F_1 < \cdots < F_K \ .$$

When a quantitative measure is available, there are other regression techniques which can be used. These are discussed in Chapter 6.

REFERENCE
Abelson and Tukey, <u>Ann. Math. Stat.</u> (1963).

1.1. Permutation Covariance Matrix

Define

$$W_{ij}^* = \sum_{\substack{i'=1 \\ (i',j') \neq (i,j)}}^{K} \sum_{j'=1}^{n_{i'}} U((X_{ij}, \delta_{ij}), (X_{i'j'}, \delta_{i'j'})) \ ,$$

and rename

$$W_{11}^*, \ \ldots, \ W_{1n_1}^*, \ \ldots, \ W_{K1}^*, \ \ldots, \ W_{Kn_K}^*$$

to

$$W_1^*, \ \ldots, \ W_N^* \ .$$

To calculate the permutation distribution of $\underset{\sim}{W}$, suppose we are given $W_1^*, \ \ldots, \ W_N^*$. Under H_0^* we sample these W_i^* without replacement, letting W_1 be the sum of the first n_1 sampled, W_2 be the sum of the next n_2 sampled, and so on.

The covariance matrix of $W = (W_1, \ldots, W_K)'$ under this sampling scheme is

$$\Sigma_{0,P}^* = \frac{1}{N} \left(\frac{\displaystyle\sum_{i=1}^{K} \sum_{j=1}^{n_i} (W_{ij}^*)^2}{N-1} \right)$$

$$\times \begin{pmatrix} n_1(N-n_1) & & & \\ & \ddots & & -n_i n_j \\ & & \ddots & \\ & -n_i n_j & & \ddots \\ & & & n_K(N-n_K) \end{pmatrix}.$$

The matrix $\Sigma_{0,P}^*$ can be used in place of $N^3 \hat{\Sigma}_0^*$, although the two are asymptotically equivalent.

REFERENCE
Marcuson and Nordbrock, <u>Biom. Zeit./Biom. J.</u> (1981).

1.2. Distribution under H_0
Under the hypothesis of interest

$$H_0 : F_1 = \cdots = F_K ,$$

Breslow shows

$$W \overset{a}{\sim} N(\underset{\sim}{\mu}_0, \ N^3 \ \underset{\sim}{\Sigma}_0) \ ,$$

where the elements of $\underset{\sim}{\Sigma}_0$ are

$$\sigma_{ij}^0 = -\lambda_i \lambda_j \int_0^\infty (1 - H_i)(1 - H_j) dH_u \quad \text{if} \quad i \ne j \ ,$$

$$\sigma_{ii}^0 = \lambda_i \int_0^\infty [(1 - H)(1 - H_i) - \lambda_i (1 - H_i)^2] dH_u \ .$$

To estimate this covariance matrix, it is easier to use the Mantel-Haenszel statistic (given next) than to substitute estimates for H, H_i, H_u.

REFERENCE
Breslow, <u>Biometrika</u> (1970).

2. GENERALIZED MANTEL-HAENSZEL TEST (TARONE AND WARE)

Let the ordered combined sample be denoted by

$$(Z_{(1)}, \ \zeta_{(1)}), \ \dots, \ (Z_{(N)}, \ \zeta_{(N)}) \ ,$$

and let

$$\Re_{(i)} = \Re(Z_{(i)}-) \ .$$

For each uncensored time point, construct a $2 \times K$ table; that is, if $(Z_{u(i)}, 1)$ is the i-th uncensored observation, form

Population

	1	2	...	K	
Dead	$a_{i1}=0$	$a_{i2}=1$	...	$a_{iK}=0$	m_{i1}
Alive			...		m_{i2}
	n_{i1}	n_{i2}	...	n_{iK}	$N_i = \#\Re_{u(i)}$

Notice that for $K = 2$, the tables are the transpose of the previous 2×2 tables in Chapter 4, Section 2.

Under $H_0 : F_1 = \cdots = F_K$,

$$E_0(\underset{\sim}{A}_i) = (E_0(A_{i1}), \ldots, E_0(A_{iK}))' ,$$

$$= \left(\frac{m_{i1}n_{i1}}{N_i}, \ldots, \frac{m_{i1}n_{iK}}{N_i} \right)' ,$$

$$\underset{\sim}{\Sigma}_0(\underset{\sim}{A}_i) = \left(\frac{m_{i1}m_{i2}}{N_i - 1} \right)$$

$$\times \begin{pmatrix} \frac{n_{i1}}{N_i}\left(1 - \frac{n_{i1}}{N_i}\right) & & -\frac{n_{ik}\,n_{i\ell}}{N_i\,N_i} \\ & \cdot & \\ -\frac{n_{ik}\,n_{i\ell}}{N_i\,N_i} & & \frac{n_{iK}}{N_i}\left(1 - \frac{n_{iK}}{N_i}\right) \end{pmatrix} .$$

Define

$$\underset{\sim}{a} - E_0(\underset{\sim}{A}) = \sum_i w_i(\underset{\sim}{a}_i - E_0(\underset{\sim}{A}_i)) \; ,$$

$$\underset{\sim}{\Sigma}_0 = \sum_i w_i^2 \; \underset{\sim}{\Sigma}_0(\underset{\sim}{A}_i) \; ,$$

where the w_i are weights.   There are three <u>special cases</u>:

   (i) $w_i \equiv 1$ gives the generalized Mantel-Haenszel test.

   (ii) $w_i = N_i$ gives the generalized Gehan test.

   (iii) $w_i = \sqrt{N_i}$ gives high efficiency over a range of alternatives.

In Section 1, we obtained the asymptotic covariance matrix of the generalized Gehan statistic under the more restrictive hypothesis H_0^*. The asymptotic covariance matrix under H_0 is $\underset{\sim}{\Sigma}_0$, and the formula given here is computationally easier than Breslow's approach.

REFERENCE
Tarone and Ware, <u>Biometrika</u> (1977).

<u>Types of Tests</u>.

1. <u>Omnibus χ^2 Test</u>.   Since $\underset{\sim}{\Sigma}_0$ is singular, delete one population, say the first.   Define $\underset{\sim}{a}_{-1} - E_0(\underset{\sim}{A}_{-1})$ and $\underset{\sim}{\Sigma}_{0,-1}$ to be $\underset{\sim}{a} - E_0(\underset{\sim}{A})$ and $\underset{\sim}{\Sigma}_0$ respectively with the first population deleted.   Then

$$W = (\underset{\sim}{a}_{-1} - E_0(\underset{\sim}{A}_{-1}))' \; \underset{\sim}{\Sigma}_{0,-1}^{-1} (\underset{\sim}{a}_{-1} - E_0(\underset{\sim}{A}_{-1})) \overset{a}{\sim} \chi_{K-1}^2$$

under H_0. The value of W will be the same no matter which population is deleted.

For the generalized Mantel-Haenszel test $(w_i \equiv 1)$, we have the approximate test

$$\sum \frac{(0 - E)^2}{E} = \sum_{k=1}^{K} \frac{(a_k - E_0(A_k))^2}{E_0(A_k)} \approx \chi^2_{K-1} \text{ ,}$$

where

$$a_k = \sum_i a_{ik} \text{ ,}$$

$$E_0(A_k) = \sum_i E_0(A_{ik}) = \sum_i \frac{m_{i1} n_{ik}}{N_i} \text{ .}$$

Although this test is somewhat conservative since $\Sigma(0 - E)^2/E \leq W$, it is simpler to use because no matrix inversion is involved.

REFERENCES
Peto and Pike, <u>Biometrics</u> (1973).
Peto et al., <u>British J. Cancer</u> (1976, 1977).

2. <u>Test for Trend</u>. Suppose

$$H_1 : F_1 < \cdots < F_K \text{ .}$$

For the choices of $\underset{\sim}{\ell}$ in Section 1, use the statistic

$$\underset{\sim}{\ell}'(\underset{\sim}{a} - E_0(\underset{\sim}{A})) \text{ ,}$$

which asymptotically has a normal distribution.

REFERENCE
Tarone, _Biometrika_ (1975).

SIX

NONPARAMETRIC METHODS: REGRESSION

1. COX PROPORTIONAL HAZARDS MODEL

Let $T_1, \ldots, T_n; C_1, \ldots, C_n$ be independent random variables. C_i is the censoring time associated with the survival time T_i. We observe $(Y_1, \delta_1), \ldots, (Y_n, \delta_n)$ where

$$Y_i = T_i \wedge C_i , \quad \delta_i = I(T_i \leq C_i) .$$

Also available are $x_{\sim 1}, \ldots, x_{\sim n}$, where

$$x_{\sim i} = (x_{\sim i1}, \ldots, x_{\sim ip})'$$

is the vector of independent variables or covariates associated with the dependent variable T_i.

Recall that the hazard function is

$$\lambda(t; \underset{\sim}{x}) = \frac{f(t; \underset{\sim}{x})}{1 - F(t; \underset{\sim}{x})} \, ,$$

where we have included the dependence of the distribution of T upon the covariates in $\underset{\sim}{x}$. The <u>proportional hazards model</u> assumes

$$\lambda(t; \underset{\sim}{x}) = e^{\underset{\sim}{\beta}' \underset{\sim}{x}} \lambda_0(t) \, ,$$

where $\underset{\sim}{\beta} = (\beta_1, \ldots, \beta_p)'$ is the vector of regression coefficients. The hazard rate is the product of a scalar and the function $\lambda_0(t)$, where the scalar depends on the regression coefficients and the covariates. The theory could work if $e^{\underset{\sim}{\beta}' \underset{\sim}{x}}$ were replaced by any sensible $h(\underset{\sim}{\beta}'x)$ where h is positive. Both the regression coefficients $\underset{\sim}{\beta}$ and the underlying hazard function $\lambda_0(t)$ are unknown.

We say that a family of distribution functions is a family of <u>Lehmann alternatives</u> if there exists a d.f. F_0 such that for any F in the family

$$1 - F = (1 - F_0)^\gamma$$

for some positive γ , or in terms of survival functions

$$S = S_0^\gamma \, .$$

The proportional hazards model implies that the distribution functions form a family of Lehmann alternatives:

$$S(t; \underset{\sim}{x}) = \exp\left\{-\int_0^t \lambda(u; \underset{\sim}{x})du\right\} ,$$

$$= \exp\left\{-e^{\underset{\sim}{\beta}'\underset{\sim}{x}} \int_0^t \lambda_0(u)du\right\} ,$$

$$= \exp\left\{-\int_0^t \lambda_0(u)du\right\}^{e^{\underset{\sim}{\beta}'\underset{\sim}{x}}} ,$$

$$= S_0(t)^{e^{\underset{\sim}{\beta}'\underset{\sim}{x}}} ,$$

where

$$S_0(t) = \exp\left\{-\int_0^t \lambda_0(u)du\right\} .$$

Consider the special case in which $p = 1$.

$$x_i = \begin{cases} 1 & \text{if the i-th observation is from} \\ & \text{Population 1 ,} \\ 0 & \text{if the i-th observation is from} \\ & \text{Population 2 .} \end{cases}$$

Then

$$e^{\beta x_i} = \begin{cases} e^\beta = \gamma & \text{if } i \text{ is from Population 1 ,} \\ 1 & \text{if } i \text{ is from Population 2 ,} \end{cases}$$

and consequently the survival functions for Population 1 and Population 2 are related by

$$S_1(t) = S_2^\gamma(t) .$$

1.1. Conditional Likelihood Analysis

Cox writes:

"Suppose then that $\lambda_0(t)$ is arbitrary. No information can be contributed about β by time intervals in which no failures occur because the component $\lambda_0(t)$ might conceivably be identically zero in such intervals. We therefore argue conditionally on the set of instants at which failures occur; in discrete time we shall condition also on the observed multiplicities. Once we require a method of analysis holding for all $\lambda_0(t)$, consideration of this conditional distribution seems inevitable."

Assume there are no ties; ties will be treated later. Order the observed times

$$y_{(1)} < y_{(2)} < \cdots < y_{(n)} ,$$

and let $\delta_{(i)}$ be the censoring indicator and $x_{(i)}$ be the covariate associated with $y_{(i)}$. Also denote $\mathbb{R}_{(i)} = \mathbb{R}(y_{(i)}-)$. For each uncensored time $y_{(i)}$,

$$P\{a \text{ death in } [y_{(i)}, y_{(i)} + \Delta y) | \mathbb{R}_{(i)}\}$$

$$\cong \sum_{j \in \mathbb{R}_{(i)}} e^{\beta' x_j} \lambda_0(y_{(i)}) \Delta y ,$$

$P\{$death of (i) at time $y_{(i)}\big|$

one death in $\Re_{(i)}$ at time $y_{(i)}\}$

$$= \frac{e^{\underset{\sim}{\beta}'\underset{\sim}{x}_{(i)}}}{\displaystyle\sum_{j\in\Re_{(i)}} e^{\underset{\sim}{\beta}'\underset{\sim}{x}_j}} \cdot$$

Taking the product of these conditional probabilities gives a (so-called) conditional likelihood:

$$L_c(\underset{\sim}{\beta}) = \prod_u \frac{e^{\underset{\sim}{\beta}'\underset{\sim}{x}_{(i)}}}{\displaystyle\sum_{j\in\Re_{(i)}} e^{\underset{\sim}{\beta}'\underset{\sim}{x}_j}} \cdot$$

Cox suggests that we <u>treat his conditional likeli-hood as an ordinary likelihood</u>. In particular, to find the maximum likelihood estimate, use the score vector and the sample information matrix:

$$\frac{\partial}{\partial \underset{\sim}{\beta}} \log L_c(\underset{\sim}{\beta}) = \left(\frac{\partial}{\partial \beta_1} \log L_c(\underset{\sim}{\beta}), \ldots, \frac{\partial}{\partial \beta_p} \log L_c(\underset{\sim}{\beta})\right)',$$

$$\underset{\sim}{i}(\underset{\sim}{\beta}) = - \frac{\partial^2}{\partial \underset{\sim}{\beta}^2} \log L_c(\underset{\sim}{\beta}),$$

$$= - \begin{pmatrix} \dfrac{\partial^2}{\partial\beta_1\partial\beta_1} \log L_c(\underset{\sim}{\beta}) & \cdots & \dfrac{\partial^2}{\partial\beta_1\partial\beta_p} \log L_c(\underset{\sim}{\beta}) \\ & \vdots & \vdots \\ \dfrac{\partial^2}{\partial\beta_p\partial\beta_1} \log L_c(\underset{\sim}{\beta}) & \cdots & \dfrac{\partial^2}{\partial\beta_p\partial\beta_p} \log L_c(\underset{\sim}{\beta}) \end{pmatrix}.$$

We want to solve the equations

$$\frac{\partial}{\partial \underset{\sim}{\beta}} \log L_c(\underset{\sim}{\beta}) = \underset{\sim}{0} \ ,$$

which usually requires iterative methods. Therefore, if $\hat{\underset{\sim}{\beta}}^0$ is an initial guess, let

$$\hat{\underset{\sim}{\beta}}^1 = \hat{\underset{\sim}{\beta}}^0 + \underset{\sim}{i}^{-1}(\hat{\underset{\sim}{\beta}}^0) \frac{\partial}{\partial \underset{\sim}{\beta}} \log L_c(\hat{\underset{\sim}{\beta}}^0) \ .$$

If $\hat{\underset{\sim}{\beta}}$ is the solution, Cox asserts

$$\hat{\underset{\sim}{\beta}} \overset{a}{\sim} N(\underset{\sim}{\beta}, \underset{\sim}{i}^{-1}(\underset{\sim}{\beta})) \ .$$

Taking derivatives of

$$\log L_c(\underset{\sim}{\beta}) = \sum_u \left[\underset{\sim}{\beta}'\underset{\sim}{x}_{(i)} - \log \left(\sum_{j \in \Re_{(i)}} e^{\underset{\sim}{\beta}'\underset{\sim}{x}_j} \right) \right] \ ,$$

we obtain formulas for the score vector and sample information matrix:

$$\frac{\partial}{\partial \beta_k} \log L_c(\underset{\sim}{\beta}) = \sum_u \left(x_{(i)k} - \frac{\sum\limits_{j \in \Re_{(i)}} x_{jk} \, e^{\underset{\sim}{\beta}'\underset{\sim}{x}_j}}{\sum\limits_{j \in \Re_{(i)}} e^{\underset{\sim}{\beta}'\underset{\sim}{x}_j}} \right) \ ,$$

$$i_{k\ell}(\underset{\sim}{\beta}) = - \frac{\partial^2}{\partial \beta_k \partial \beta_\ell} \log L_c(\underset{\sim}{\beta}) \ ,$$

$$
= \sum_u \left(\frac{\sum\limits_{j \in \mathbb{R}_{(i)}} x_{jk} \, x_{j\ell} \, e^{\underset{\sim}{\beta}' \underset{\sim}{x}_j}}{\sum\limits_{j \in \mathbb{R}_{(i)}} e^{\underset{\sim}{\beta}' \underset{\sim}{x}_j}} \right.
$$

$$
\left. - \frac{\sum\limits_{j \in \mathbb{R}_{(i)}} x_{jk} \, e^{\underset{\sim}{\beta}' \underset{\sim}{x}_j}}{\sum\limits_{j \in \mathbb{R}_{(i)}} e^{\underset{\sim}{\beta}' \underset{\sim}{x}_j}} \times \frac{\sum\limits_{j \in \mathbb{R}_{(i)}} x_{j\ell} \, e^{\underset{\sim}{\beta}' \underset{\sim}{x}_j}}{\sum\limits_{j \in \mathbb{R}_{(i)}} e^{\underset{\sim}{\beta}' \underset{\sim}{x}_j}} \right) .
$$

For testing $H_0 : \underset{\sim}{\beta} = \underset{\sim}{0}$, Cox uses the Rao-type statistic

$$
\left(\frac{\partial}{\partial \underset{\sim}{\beta}} \log L_c(\underset{\sim}{0}) \right)' \, \underset{\sim}{i}^{-1}(\underset{\sim}{0}) \left(\frac{\partial}{\partial \underset{\sim}{\beta}} \log L_c(\underset{\sim}{0}) \right) ,
$$

which is asymptotically χ^2_p under H_0. The score vector and sample information matrix have simpler forms at $\underset{\sim}{\beta} = \underset{\sim}{0}$:

$$
\frac{\partial}{\partial \beta_k} \log L_c(\underset{\sim}{0}) = \sum_u (x_{(i)k} - \bar{x}_{(i)k}) ,
$$

$$
i_{k\ell}(\underset{\sim}{0}) = \sum_u \left(\frac{1}{n_i} \sum_{j \in \mathbb{R}_{(i)}} x_{jk} \, x_{j\ell} \right.
$$

$$
\left. - \bar{x}_{(i)k} \, \bar{x}_{(i)\ell} \right) ,
$$

$$= \sum_u \left(\frac{1}{n_i} \sum_{j \in \mathcal{R}_{(i)}} (x_{jk} - \bar{x}_{(i)k}) \right.$$

$$\left. \times \, (x_{j\ell} - \bar{x}_{(i)\ell}) \right) ,$$

where

$$\bar{x}_{\sim(i)} = \frac{\sum\limits_{j \in \mathcal{R}_{(i)}} x_{\sim j}}{n_i} , \quad n_i = \# \text{ in } \mathcal{R}_{(i)} .$$

The sample information matrix is simply a sum of the covariate covariance matrices for the risk sets of the uncensored observations.

Consider the <u>special case</u> $p = 1$ and

$$x_i = \begin{cases} 1 & \text{if } i \text{ is in sample 1 ,} \\ 0 & \text{if } i \text{ is in sample 2 .} \end{cases}$$

Then using the notation for the Mantel–Haenszel test,

$$\frac{\partial}{\partial \beta} \log L_c(0) = \sum_u \{x_{(i)} - \bar{x}_{(i)}\} = \sum_u \left(a_i - \frac{n_{i1}}{n_i} \right) ,$$

$$i(0) = \sum_u \left(\frac{1}{n_i} \sum_{j \in \mathcal{R}_{(i)}} x_j^2 - \bar{x}_{(i)}^2 \right) ,$$

$$= \sum_u \frac{n_{i1}}{n_i} \left(1 - \frac{n_{i1}}{n_i} \right) .$$

Therefore, in the case of no ties, Cox's test is exactly equal to the Mantel–Haenszel test.

REFERENCES

Cox, JRSS B (1972).

Prentice and Kalbfleisch, Biometrics (1979), has a
   nice survey of the Cox procedure.

Kalbfleisch and Prentice, The Statistical Analysis
   of Failure Time Data (1980), is an excellent
   new text on the Cox approach.

1.2. Justification of the Conditional Likelihood

Marginal Likelihood for Ranks. Make the crucial
assumption of no ties. When ties are present, the
following argument breaks down.

Suppose the data are uncensored. Let $Y_1, \ldots, Y_n$
be independent and Y_i have d.f. F_i with density
f_i. Denote

$$Y = (Y_1, \ldots, Y_n) ,$$

$$R = (R_1, \ldots, R_n) ,$$

where

$$R_i = \text{rank of } Y_i .$$

Then the probability for the rank vector r is

$$p(r) = \int \cdots \int_{u_1 < \cdots < u_n} \prod_{i=1}^{n} f_{(i)}(u_i) \, du_1 \cdots du_n ,$$

where $f_{(i)}$ is the density corresponding to $y_{(i)}$.
For example, if $n = 3$ and $r = (3, 1, 2)$,

$$p(r) = P\{R_1 = 3, R_2 = 1, R_3 = 2\} ,$$

$$= \int \int \int_{u_1 < u_2 < u_3} f_2(u_1) \, f_3(u_2) \, f_1(u_3) \, du_1 du_2 du_3 \; .$$

Kalbfleisch and Prentice show that when

$$F_i(t) = 1 - \exp\left(-e^{\beta' x_i} \int_0^t \lambda_0(u) du\right) ,$$

then

$$p(r) = \prod_{i=1}^{n} \frac{e^{\beta' x_{(i)}}}{\sum_{j \in \mathcal{R}_{(i)}} e^{\beta' x_j}} \; .$$

Now allow censoring. Use the usual notation $(Y_1, \delta_1), \ldots, (Y_n, \delta_n)$, and let Y_i have d.f. F_i and density f_i. Define the rank vector

$$R^{u/c} = (R_1^{u/c}, \ldots, R_n^{u/c}) ,$$

where

$$R_i^{u/c} = \begin{cases} \text{rank of } Y_i \text{ among uncensored} \\ \quad \text{observations if } \delta_i = 1 , \\[2mm] \text{rank of the preceding uncensored} \\ \quad \text{observation if } \delta_i = 0 , \end{cases}$$

and the indicator vector

$$\delta = (\delta_1, \ldots, \delta_n) \; .$$

Then the probability of $(r^{u/c}, \delta)$ is

$$p(r^{u/c}, \delta) = \int \cdots \int_{u_1 < \cdots < u_{n_u}} \prod_{i=1}^{n_u} \left\{ f_{u(i)}(u_i) \right.$$

$$\times \prod_{j \in C_{i,i+1}} [1 - F_j(u_i)] \left. \right\} du_1 \cdots du_{n_u} ,$$

where $f_{u(i)}$ is the density corresponding to the
i-th ordered uncensored observation, $C_{i,i+1}$ is the
set of indices corresponding to the censored observations between the i-th and $(i+1)$-th ordered uncensored observations, and n_u is the total number of
uncensored observations.

For example, for $r = (2, 1, 1)$ and $\delta = (1, 1, 0)$
corresponding to the picture

$$\frac{\qquad X \quad 0 \quad X \qquad}{Y_2 \quad Y_3 \quad Y_1} ,$$

the rank probability is

$$p((2,1,1), (1,1,0)) = \int \int_{u_1 < u_2} f_2(u_1) [1 - F_3(u_1)]$$

$$\times f_1(u_2) \, du_1 du_2 .$$

Kalbfleisch and Prentice show that if

$$F_i(t) = 1 - \exp\left(-e^{\beta' x_i} \int_0^t \lambda_0(u)du\right) ,$$

then

$$p(r^{u/c}, \delta) = \prod_u \frac{e^{\beta' x_{(i)}}}{\sum_{j \in \mathfrak{R}_{(i)}} e^{\beta' x_j}} = L_c .$$

REFERENCE
Kalbfleisch and Prentice, _Biometrika_ (1973).

Partial Likelihood. Consider the sequence of
pairs of random quantities

$$(X_1, S_1; X_2, S_2; \ldots; X_m, S_m) .$$

In regression with censored data let $y_{u(i)}$ denote
the i-th ordered uncensored observation. Think of
X_i as containing all the censoring information in
$[y_{u(i-1)}, y_{u(i)})$ together with the information that
a failure occurs at time $y_{u(i)}$, and think of S_i
as containing the information that the particular individual with covariate $x_{u(i)}$ failed at time $y_{u(i)}$.
The marginal likelihood of $S_1, \ldots, S_m$ is

$$p(S_1, \ldots, S_m | \beta) = \prod_{i=1}^m p(S_i | S_1, \ldots, S_{i-1}; \beta) ,$$

and the conditional likelihood of $S_1, \ldots, S_m$ given
$X_1, \ldots, X_m$ is

$$p(S_1, \ldots, S_m | X_1, \ldots, X_m; \underset{\sim}{\beta})$$

$$= \prod_{i=1}^{m} p(S_i | S_1, \ldots, S_{i-1}; X_1, \ldots, X_m; \underset{\sim}{\beta}) \; .$$

The full likelihood is

$$p(X_1, \ldots, X_m; S_1, \ldots, S_m | \underset{\sim}{\beta})$$

$$= \prod_{i=1}^{m} p(X_i, S_i | X_1, \ldots, X_{i-1}; S_1, \ldots, S_{i-1}; \underset{\sim}{\beta}) \; ,$$

$$= \prod_{i=1}^{m} p(X_i | X_1, \ldots, X_{i-1}; S_1, \ldots, S_{i-1}; \underset{\sim}{\beta})$$

$$\times \prod_{i=1}^{m} p(S_i | X_1, \ldots, X_{i-1}, X_i; S_1, \ldots, S_{i-1}; \underset{\sim}{\beta}) \; .$$

Cox calls the second product, that is,

$$\prod_{i=1}^{m} p(S_i | X_1, \ldots, X_{i-1}, X_i; S_1, \ldots, S_{i-1}; \underset{\sim}{\beta}) \; ,$$

the partial likelihood.

In regression with censored data the partial likelihood coincides with L_c , which we have called a conditional likelihood. A comparison of the definitions of partial likelihood and conditional likelihood shows that the partial likelihood is not a true conditional likelihood, nor is it a marginal likelihood.

Cox claims that the partial likelihood contains most of the information about $\underset{\sim}{\beta}$ for regression with censored data and that we can ignore the first product, that is,

$$\prod_{i=1}^{m} p(X_i | X_1, \ldots, X_{i-1}; S_1, \ldots, S_{i-1}; \beta) ,$$

without losing much. Efron and Oakes have compared
the Fisher information in the partial likelihood to
the Fisher information in the full likelihood for a
variety of models. Usually the information in L_c
is very high with efficiency $\geq 90\%$, and in rare cases,
L_c even carries as much information as the full
likelihood.

REFERENCES
Cox, Biometrika (1975).
Efron, JASA (1977).
Oakes, Biometrika (1977).

1.3. Justification of Asymptotic Normality

In his 1972 paper Cox asserts that $\hat{\beta}$, the solution to

$$\frac{\partial}{\partial \beta} \log L_c(\beta) = 0 ,$$

is asymptotically normally distributed. In his 1975
paper Cox gives a heuristic argument which is similar
to the standard maximum likelihood argument.

Tsiatis gives a proof of the asymptotic normality
of $\hat{\beta}$ using integral representations and stochastic
processes which is similar to the proof given by
Breslow and Crowley of the asymptotic normality of
the PL estimator and the proof given by Crowley for
the Mantel-Haenszel statistic.

Bailey gives an argument using Hajek projections.

REFERENCES
Cox, Biometrika (1975).

Bailey, Univ. of Chicago thesis (1979).
Tsiatis, <u>Ann. Stat.</u> (1981).

1.4. Estimation of $S(t; \underset{\sim}{x})$

Under the Cox proportional hazards model,

$$S(t; \underset{\sim}{x}) = \exp\left(-e^{\underset{\sim}{\beta}'\underset{\sim}{x}} \int_0^t \lambda_0(u)du\right) ,$$

$$= \exp\left(-e^{\underset{\sim}{\beta}'\underset{\sim}{x}} \Lambda_0(t)\right) ,$$

$$= S_0(t)^{e^{\underset{\sim}{\beta}'\underset{\sim}{x}}} ,$$

where

$$S_0(t) = e^{-\Lambda_0(t)} .$$

To estimate $S(t; x)$, we substitute $\hat{\underset{\sim}{\beta}}$ for $\underset{\sim}{\beta}$, but how do we estimate $\Lambda_0(t)$ or $S_0(t)$?

Breslow assumes $\lambda_0(t)$ is constant between uncensored observations:

$$\hat{\lambda}_{0,B}(t) = \cfrac{1}{(y_{u(i)} - y_{u(i-1)}) \sum\limits_{j \in \mathcal{R}_{u(i)}} e^{\underset{\sim}{\beta}'\underset{\sim}{x}_j}}$$

$$\text{if } y_{u(i-1)} < t < y_{u(i)} .$$

However, he estimates $S_0(t)$ by

$$\hat{S}_{0,B}(t) = \prod_{y_{(i)} \leq t} \left(1 - \frac{\delta_{(i)}}{\sum_{j \in \mathscr{R}_{(i)}} e^{\hat{\beta}' x_j}} \right).$$

Notice that $\hat{\Lambda}_{0,B}(t) = \int_0^t \hat{\lambda}_{0,B}(u)du$ and $\hat{S}_{0,B}(t)$ are inconsistent in the sense that

$$\hat{S}_{0,B}(t) \neq e^{-\hat{\Lambda}_0(t)}$$

even for $t = y_{(i)}$. Moreover, $\hat{S}_{0,B}(t)$ can take ne-gative values.

Tsiatis uses

$$\hat{S}_{0,T}(t) = e^{-\hat{\Lambda}_{0,T}(t)},$$

where

$$\hat{\Lambda}_{0,T}(t) = \sum_{y_{(i)} \leq t} \frac{\delta_{(i)}}{\sum_{j \in \mathscr{R}_{(i)}} e^{\hat{\beta}' x_j}},$$

but $\hat{S}_{0,T}$ does not simplify to the PL estimator when $\hat{\beta} = 0$. Notice that $\hat{\Lambda}_0(t)$ is a step function.

Link uses a linear smooth of $\hat{\Lambda}_{0,T}(t)$, which is the integral of the Breslow estimate of $\lambda_0(t)$, and, like Tsiatis, defines

$$\hat{S}_{0,L}(t) = e^{-\hat{\Lambda}_{0,L}(t)} .$$

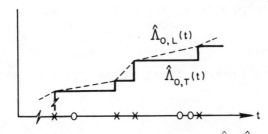

Both Tsiatis and Link calculate $\widehat{Var}(\hat{S}_0(t))$,
Tsiatis using a likelihood model and Link using the
delta method. Link also uses Monte Carlo methods to
study the confidence intervals associated with

$$\hat{S}_{0,L}(t), \log \hat{S}_{0,L}(t) , \quad \text{and}$$

$$\text{logit } \hat{S}_{0,L}(t) = \log \frac{\hat{S}_{0,L}(t)}{1 - \hat{S}_{0,L}(t)} ,$$

and finds that the coverage probabilities with
$\hat{S}_{0,L}(t)$ tend to be too low, those with
$\text{logit } \hat{S}_{0,L}(t)$ too high, and those with $\log \hat{S}_{0,L}(t)$
approximately correct. These results concerning the
confidence interval coverages hold also for the PL
estimator.
    Alternative estimators of $S(t; \underset{\sim}{x})$, which are
computationally more complicated, have been proposed
by Cox, Efron, and Kalbfleisch-Prentice.

REFERENCES
Breslow, JRSS B (1972), in Discussion on Cox's
    paper.
_____, Biometrics (1974).
Tsiatis, Univ. Wisconsin Tech. Report No. 524
    (1978).
_____, Ann. Stat. (1981).
Link, Stanford Univ. Tech. Report No. 45 (1979).

1.5. Discrete or Grouped Data

Denote the ordered distinct survival times by

$$y'_{(1)} < \cdots < y'_{(r)} \; ,$$

and let

$$\mathcal{R}_{(i)} = \text{risk set at time } \; y'_{(i)}{}^{-} \; ,$$

$$\mathcal{D}_{(i)} = \text{death set at time } \; y'_{(i)} \; , \text{ that is, the}$$
$$\text{set of individuals who die at time } y'_{(i)} \, ,$$

$$d_i = \#(\mathcal{D}_{(i)}) \; .$$

Cox suggests using

$$L_c = \prod_{i=1}^{r} P\{\mathcal{D}_{(i)} \mid \mathcal{R}_{(i)}, \, d_i\} \; ,$$

with

$$P\{\mathcal{D}_{(i)} \mid \mathcal{R}_{(i)}, d_i\} = \frac{\exp\left(\sum_{j \in \mathcal{D}_{(i)}} \beta' x_j\right)}{\sum_{\mathcal{D}^*_{(i)}} \exp\left(\sum_{j \in \mathcal{D}^*_{(i)}} \beta' x_j\right)},$$

where the summation in the denominator is over all subsets $\mathcal{D}^*_{(i)} \subset \mathcal{R}_{(i)}$ such that $\#(\mathcal{D}^*_{(i)}) = d_i$. For $i = 1, \ldots, r$, there are $\binom{n_i}{d_i}$ subsets to consider so for even modest-sized data sets, this approach is not computationally feasible.

An alternative likelihood, proposed by Peto, Breslow, and Kalbfleisch-Prentice, is

$$L_c = \prod_{i=1}^{r} \frac{\exp\left(\sum_{j \in \mathcal{D}_{(i)}} \beta' x_j\right)}{\left(\sum_{j \in \mathcal{R}_{(i)}} e^{\beta' x_j}\right)^{d_i}},$$

which seems to work reasonably well when the number of ties is not excessive.

Neither of these likelihoods strictly adheres to a Lehmann alternative model or a proportional hazards model, but the next proposal by Prentice and Gloeckler does.

Prentice and Gloeckler assume that the time axis is partitioned by

$$0 = a_0 < a_1 < \cdots < a_{r-1} < a_r = \infty,$$

and

$$A_j = [a_{j-1}, a_j) \; .$$

If a survival time falls in the interval A_j , then record time j. Denote

$$\alpha_j = \exp\left(-\int_{a_{j-1}}^{a_j} \lambda_0(t)dt\right) ,$$

which is the conditional probability of an individual with covariate $\underset{\sim}{x} = \underset{\sim}{0}$ surviving A_j , given that he has survived A_{j-1}. Then the probability of the i-th observation surviving to the beginning of A_j is

$$\prod_{k=1}^{j-1} \alpha_k^{e^{\beta' x_i}} ,$$

and

$$P\{Y_i = j, \; \delta_i\} = \left(\prod_{k=1}^{j-1} \alpha_k^{e^{\beta' x_i}}\right)\left(1 - \alpha_j^{e^{\beta' x_i}}\right)^{\delta_i} .$$

The full likelihood is

$$L = \prod_{i=1}^{n} P\{Y_i = j, \; \delta_i\} ,$$

which is a function of the unknown parameters $\underset{\sim}{\beta}, \alpha_1, \ldots, \alpha_r$.

To estimate these parameters use maximum likelihood. Notice that the α_j's are restricted by

$$0 < \alpha_j < 1, \quad j = 1, \ldots, r \ , \quad \text{and} \quad \sum_{j=1}^{r} \alpha_j = 1 \ .$$

Eliminate α_r and let

$$\gamma_j = \log(-\log \alpha_j) \ , \quad j = 1, \ldots, r-1 \ ,$$

so that

$$-\infty < \gamma_j < +\infty \ , \quad j = 1, \ldots, r-1 \ .$$

Maximizing with respect to $\gamma_1, \ldots, \gamma_{r-1}$ is simpler
than maximizing with respect to $\alpha_1, \ldots, \alpha_r$ because
there is no need to worry about the boundaries. Also,
Newton-Raphson convergence is faster.

REFERENCES
Cox, JRSS B (1972).
Kalbfleisch and Prentice, JRSS B (1972), and
Peto, JRSS B (1972), in Discussion on Cox's paper.
Breslow, Biometrics (1974).
Prentice and Gloeckler, Biometrics (1978).

1.6. Time Dependent Covariates

We generalize to the situation in which the covariate is allowed to vary with time. Therefore, together with

$$Y_i = T_i \wedge C_i \ , \quad \delta_i = I(T_i \leq C_i) \ ,$$

we observe $\underset{\sim}{x}_i(t)$. The proportional hazards model

assumes the hazard function of the i-th observation
to be

$$\lambda_i(t) = e^{\beta' x_i(t)} \lambda_0(t) \ ,$$

so that

P{death of (i) at time $y_{(i)}$ |

   one death in $\Re_{(i)}$ at time $y_{(i)}$}

$$= \frac{e^{\beta' x_{(i)}(y_{(i)})}}{\sum\limits_{j \in \Re_{(i)}} e^{\beta' x_j(y_{(i)})}} \ ,$$

and the conditional likelihood becomes

$$L_c = \prod_u \frac{e^{\beta' x_{(i)}(y_{(i)})}}{\sum\limits_{j \in \Re_{(i)}} e^{\beta' x_j(y_{(i)})}} \ .$$

In the time varying case, no proof exists for the
asymptotic normality of $\hat{\beta}$. Also, for moderate to
large data sets the computations become slow and ex-
pensive.

1.7.   Example 1.   Stanford Heart Transplant Data

Do heart transplant patients survive longer than
heart disease patients who do not receive heart trans-
plants? Typically, a patient enters the study and
receives a transplant when a donor heart becomes
available. Upon transplantation, we say that the

patient has migrated from the no-transplant population to the transplant population, and the covariate
that indicates transplant changes from 0 to 1. Other
covariates measured include age, waiting time to
transplantation, calendar time from beginning of
study, and a mismatch score which measures the degree
of dissimilarity between donor and recipient tissues.

REFERENCES
Turnbull, Brown, and Hu, JASA (1974).
Crowley and Hu, JASA (1977).

1.8. Example 2. Adoption and Pregnancy

Are couples with an infertility problem who have
adopted a child more likely to conceive than couples
who have not adopted? Here, couples may migrate from
the childless population to the adopted population.
Censoring occurs when a couple stops trying for a
pregnancy.

REFERENCES
Lamb and Leurgans, Amer. J. Obstet. Gyn. (1979).
Leurgans, Stanford Univ. Tech. Report No. 57
    (1980).

2. LINEAR MODELS

The standard linear model is

$$T_i = \alpha + \beta x_i + e_i \; ,$$

or

$$T_i = \alpha + \underset{\sim}{\beta}' \underset{\sim}{x}_i + e_i \; , \quad i = 1, \ldots, n \; ,$$

where $e_1, \ldots, e_n$ are iid with common distribution
F. Let $C_1, \ldots, C_n$ be independent; C_i is the

censoring time associated with T_i. We observe

$$Y_i = T_i \wedge C_i \ , \quad \delta_i = I(T_i \leq C_i) \ .$$

Accelerated Time Models. Linear models are connected to hazard models through accelerated time models. Suppose Z_0 is a survival time with hazard rate

$$\lambda_0(z) = \frac{f_0(z)}{1 - F_0(z)} \ ,$$

and assume that the survival time of an individual with covariate $\underset{\sim}{x}$ has the same distribution as

$$Z_{\underset{\sim}{x}} = e^{\underset{\sim}{\beta}' \underset{\sim}{x}} Z_0 \ .$$

Notice that if $\underset{\sim}{\beta}'\underset{\sim}{x} < 0$, then $Z_{\underset{\sim}{x}}$ is shorter than Z_0 , and we say that the covariate accelerates the time to failure. The hazard rate of $Z_{\underset{\sim}{x}}$ is

$$\lambda_{\underset{\sim}{x}}(z) = \frac{f_{\underset{\sim}{x}}(z)}{1 - F_{\underset{\sim}{x}}(z)} \ ,$$

$$= \frac{f_0(e^{-\underset{\sim}{\beta}'\underset{\sim}{x}} z)e^{-\underset{\sim}{\beta}'\underset{\sim}{x}}}{1 - F_0(e^{-\underset{\sim}{\beta}'\underset{\sim}{x}} z)} \ ,$$

$$= \lambda_0(e^{-\underset{\sim}{\beta}'\underset{\sim}{x}} z)e^{-\underset{\sim}{\beta}'\underset{\sim}{x}} \ .$$

Define

$$T_{\underset{\sim}{x}} = \log Z_{\underset{\sim}{x}} .$$

Then

$$E(T_{\underset{\sim}{x}}) = \underset{\sim}{\beta}'\underset{\sim}{x} + E(\log Z_0) ,$$

$$= \underset{\sim}{\beta}'\underset{\sim}{x} + \alpha ,$$

so the accelerated time model coincides with a log-linear model:

$$T_{\underset{\sim}{x}} = \alpha + \underset{\sim}{\beta}'\underset{\sim}{x} + e , \quad \text{where}$$

$$e = \log Z_0 - E(\log Z_0) .$$

In applying linear model methods to survival data it is frequently necessary to transform the data by a logarithmic transformation in order to symmetrize it so the accelerated time model is very relevant in this regard.

REFERENCES
Prentice and Kalbfleisch, Biometrics (1979), and Kalbfleisch and Prentice, The Statistical Analysis of Failure Time Data (1980), both discuss the accelerated time model.

2.1. Linear Rank Tests

With no censoring present, the locally most powerful rank statistic for testing $H_0 : \beta = 0$ against $H_1 : \beta \neq 0$ in the case $p = 1$ is

$$\left. \frac{d}{d\beta} \log p(\underset{\sim}{r}) \right|_{\beta=0} ,$$

where $p(\underset{\sim}{r})$ is the probability of the rank vector $\underset{\sim}{r}$. Similarly, when censoring is present, use the statistic

$$\frac{d}{d\beta} \log p(\underset{\sim}{r}^{u/c}, \delta) \Big|_{\beta=0} \quad ,$$

where from Section 1

$$p(\underset{\sim}{r}^{u/c}, \delta) = \int \cdots \int\limits_{u_1 < \cdots < u_{n_u}} \prod_{i=1}^{n_u} \left\{ f_{u(i)}(u_i) \right.$$

$$\times \prod_{j \in C_{i,i+1}} [1 - F_j(u_i)] \left. \right\} du_1 \cdots du_{n_u} \quad ,$$

$$f_i(u) = f(u - \beta x_i) \quad .$$

It can be shown that

$$\frac{d}{d\beta} \log p(\underset{\sim}{r}^{u/c}, \delta) \Big|_{\beta=0}$$

$$= \sum_{i=1}^{n_u} \left\{ x_{u(i)} c_i + \left(\sum_{j \in C_{i,i+1}} x_j \right) c_i \right\} \quad ,$$

where

$$c_i = \left(\prod_{j=1}^{n_u} n_{u(j)} \right) \int_{u_1 < \cdots < u_{n_u}} \cdots \int \left\{ -\frac{d}{du_i} \log f(u_i) \right\}$$

$$\times \prod_{j=1}^{n_u} \left\{ f(u_j) [1-F(u_j)]^{m_{u(j)}} \right\} du_1 \cdots du_{n_u} \, ,$$

$$C_i = \left(\prod_{j=1}^{n_u} n_{u(j)} \right) \int_{u_1 < \cdots < u_{n_u}} \cdots \int \left\{ -\frac{d}{du_i} \log [1-F(u_i)] \right\}$$

$$\times \prod_{j=1}^{n_u} \left\{ f(u_j) [1-F(u_j)]^{m_{u(j)}} \right\} du_1 \cdots du_{n_u} \, ,$$

$$m_{u(j)} = \# \text{ in } C_{j,j+1} \, .$$

Letting the error distribution be the extreme value distribution

$$1 - F(t) = e^{-e^t} \, , \quad f(t) = e^{t-e^t} \, ,$$

(see Chapter 2, Section 2.2.), then

$$c_i = \sum_{j=1}^{i} \frac{1}{n_{u(j)}} - 1 \, ,$$

$$C_i = \sum_{j=1}^{i} \frac{1}{n_{u(j)}} \, ,$$

so the locally most powerful rank statistic in this

case becomes

$$-\sum_u (x_{(i)} - \bar{x}_{(i)}) \; ,$$

which is the numerator of the Cox statistic for testing $H_0 : \beta = 0$ (see Section 1). Peto and Peto named this the <u>log rank test</u>.

REFERENCES
Peto and Peto, <u>JRSS A</u> (1972), introduce linear
    rank tests and coin the term "log rank test".
Latta, <u>Biometrika</u> (1977), establishes a connection
    between linear rank tests and Efron's test.
Morton, <u>Biometrika</u> (1978), discusses permutation
    theory for linear rank tests.
Prentice, <u>Biometrika</u> (1978), gives the preceding
    derivation of the linear rank test statistic
    and calculates its variance.
Kalbfleisch and Prentice, <u>The Statistical Analysis</u>
    <u>of Failure Time Data</u> (1980), Chapter 6.

2.2. Least Squares Estimators

We assume here for simplicity $p = 1$, that is,

$$E(T_i) = \alpha + \beta x_i \; .$$

The estimators can be generalized to handle more than one covariate.

<u>Miller Estimators.</u> With no censoring present, the estimates $\hat{\alpha}$ and $\hat{\beta}$ minimize

$$\sum_{i=1}^{n} (y_i - \alpha - \beta x_i)^2 = n \int_{-\infty}^{\infty} z^2 \, dF_n(z) \; ,$$

where F_n is the empirical distribution function of $z_1, \ldots, z_n$ and

$$z_i = y_i - \alpha - \beta x_i .$$

With censoring present, Miller proposes to minimize

$$n \int_{-\infty}^{\infty} z^2 \, d\hat{F}(z) = \sum_{i=1}^{n} \hat{w}_i(\beta)(y_i - \alpha - \beta x_i)^2 ,$$

where $\hat{F}$ is the PL estimator based on $(z_1, \delta_1), \ldots,$ (z_n, δ_n) , and the weights $\hat{w}_1(\beta), \ldots, \hat{w}_n(\beta)$ are the jumps of the PL estimator. Notice that if $\delta_i = 0$, corresponding to a censored observation, $\hat{w}_i(\beta) = 0$, so at first glance the weighted sum of squares does not depend on the censored observations. However, the PL estimator, and therefore each weight, does depend on the censored observations.

If $\delta_{(n)} = 0$ so that the last ordered observation is censored, change it to be uncensored. Then, $\sum_1^n \hat{w}_i(\beta) = 1$.

We have written the weights as functions of β only. Since adding a constant α to each T_i results only in a shift of the PL estimator, the jumps of the PL estimator, and therefore the weights, do not depend on α.

To calculate $\hat{\alpha}, \hat{\beta}$, we differentiate with respect to α to obtain

$$\hat{\alpha} = \sum_{i=1}^{n} \hat{w}_i(\beta) y_i - \beta \sum_{i=1}^{n} \hat{w}_i(\beta) x_i .$$

Substituting this expression into the weighted sum of squares results in a function of β alone:

$$f(\beta) = \Sigma \; \hat{w}_i(\beta)(y_i - \hat{\alpha} - \beta x_i)^2 \; ,$$

which can be minimized by a search procedure.

Since the function $f(\beta)$ is not continuous, the search for the minimum can be tedious, especially in higher dimensions. As an alternative procedure Miller suggests the following modified approach. Define the initial estimate

$$\hat{\beta}^0 = \frac{\sum\limits_{u} y_i(x_i - \bar{x}_u)}{\sum\limits_{u}(x_i - \bar{x}_u)^2} \; ,$$

which is the slope of the least-squares line through the uncensored observations. With this guess $\hat{\beta}^0$, form

$$\hat{z}_i^0 = y_i - \hat{\beta}^0 x_i \; , \quad i = 1, \; \ldots, \; n \; .$$

Let $\hat{F}^0$ be the PL estimator based on $(\hat{z}_1^0, \; \delta_1), \; \ldots,$ $(\hat{z}_n^0, \; \delta_n)$, and let $\hat{w}_1(\hat{\beta}^0), \; \ldots, \; \hat{w}_n(\hat{\beta}^0)$ be the jumps of $\hat{F}^0$. Now define the new estimate

$$\hat{\beta}^1 = \frac{\sum\limits_{u} \hat{w}_i^*(\hat{\beta}^0) \; y_i(x_i - \bar{x}_u^*)}{\sum\limits_{u} \hat{w}_i^*(\hat{\beta}^0)(x_i - \bar{x}_u^*)^2} \; ,$$

where

$$\hat{w}_i^*(\hat{\beta}^0) = \frac{\hat{w}_i(\hat{\beta}^0)}{\sum\limits_u \hat{w}_i(\hat{\beta}^0)} \, ,$$

$$\bar{x}_u^* = \sum\limits_u \hat{w}_i^*(\hat{\beta}^0)x_i \, .$$

Using the renormalized weights $\hat{w}_i^*(\hat{\beta}^0)$ allows us to ignore whether the last ordered $\hat{z}_i^0$ is censored or not. Only the uncensored observations appear in the summation. The usual procedure of redefining the last ordered $\hat{z}_i^0$ to be uncensored if it is censored gives less stable results in iteratively estimating β , but it should still be used in estimating α.

We iterate the above procedure and hope for convergence. However, the sequence of estimates of β may become trapped in a loop where they oscillate between two values, in which case we take the average of the two values.

Assuming that the variability due to the weights $\hat{w}_i^*(\hat{\beta})$ is negligible,

$$\text{Var}(\hat{\beta}) = \frac{\sum\limits_u \hat{w}_i^*(\hat{\beta}) \, (y_i - \hat{\alpha} - \hat{\beta}x_i)^2}{\sum\limits_u \hat{w}_i^*(\hat{\beta}) \, (x_i - \bar{x}_u^*)^2} \, .$$

The derivation of this variance estimate, as well as the proof of the consistency of the estimates $\hat{\alpha}$ and $\hat{\beta}$, depends on the assumption that the censoring distribution of the i-th observation is

$$G_{x_i}(c) = G_0(c - \beta x_i) \, ,$$

for some distribution function G_0. If G_0 has density g_0 as pictured below, then G_{x_i} has density g_{x_i}, which is g_0 translated by βx_i:

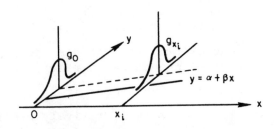

REFERENCE
Miller, _Biometrika_ (1976).

Buckley-James Estimator.   The model assumes

$$E(T_i) = \alpha + \beta x_i \ .$$

Unfortunately, we cannot observe T_i, but only Y_i, and

$$E(Y_i) \ne \alpha + \beta x_i \ .$$

Buckley and James define the pseudo random variables

$$Y_i^* = Y_i \delta_i + E(T_i | T_i > Y_i)(1 - \delta_i) \ ,$$

and calculate

$$E(Y_i^*) = \int_0^\infty u(1 - G_i(u))dF_i(u)$$

$$+ \int_0^\infty \left[\int_u^\infty \frac{s\ dF_i(s)}{1 - F_i(u)} \right] (1 - F_i(u))dG_i(u) \ ,$$

$$= \int_0^\infty u(1 - G_i(u))dF_i(u)$$

$$+ \int_0^\infty \left[\int_0^s dG_i(u) \right] sdF_i(s) \ ,$$

$$= \int_0^\infty u(1 - G_i(u))dF_i(u) + \int_0^\infty G_i(s)s\ dF_i(s) \ ,$$

$$= \int_0^\infty u\ dF_i(u) \ ,$$

$$= \alpha + \beta x_i \ .$$

Therefore, if we could observe $y_1^*, \ldots, y_n^*$, it would be reasonable to use

$$\hat{\alpha} = \bar{y}^* - \hat{\beta}\bar{x} \quad \text{and} \quad \hat{\beta} = \frac{\sum_{i=1}^n y_i^*(x_i - \bar{x})}{\sum_{i=1}^n (x_i - \bar{x})^2} \ . \tag{15}$$

Since we cannot observe all of $y_1^*, \ldots, y_n^*$, we substitute estimates for those we cannot observe. If $\delta_i = 0$, define

$$\hat{E}(T_i | T_i > y_i) = \hat{\beta} x_i + \frac{\sum\limits_{\hat{z}_k > \hat{z}_i} \hat{w}_k(\hat{\beta}) \hat{z}_k}{1 - \hat{F}(\hat{z}_i)} , \qquad (16)$$

where $\hat{z}_i = y_i - \hat{\beta} x_i$, $\hat{F}$ is the PL estimator based on $(\hat{z}_1, \delta_1), \ldots, (\hat{z}_n, \delta_n)$, and $\hat{w}_1(\hat{\beta}), \ldots, \hat{w}_n(\hat{\beta})$ are the jumps of $\hat{F}$. Then define

$$\hat{y}_i^* = y_i \delta_i + \left[\hat{\beta} x_i + \frac{\sum\limits_{\hat{z}_k > \hat{z}_i} \hat{w}_k(\hat{\beta}) \hat{z}_k}{1 - \hat{F}(\hat{z}_i)} \right] (1 - \delta_i) . \quad (17)$$

Since equations (15) give $\hat{\beta}$ as a function of y_i^* and equation (17) gives y_i^* as a function of $\hat{\beta}$, we need to iterate. As with the Miller estimate, the sequence of estimates of β may eventually oscillate between two values, and again we take the solution to be the average.

Buckley and James claim that if the estimates of β oscillate, then the difference between their two values is smaller than that for the Miller estimate. Furthermore, the validity of their method does not depend on assumptions on the censoring distributions G_i.

Buckley and James give the variance estimate

$$\hat{Var}(\hat{\beta}) = \frac{\hat{\sigma}_u^2}{\sum\limits_u (x_i - \bar{x}_u)^2} ,$$

where

$$\hat{\sigma}_u^2 = \frac{1}{n_u - 2} \sum_u (y_i - \bar{y}_u - \hat{\beta}(x_i - \bar{x}_u))^2 \, ,$$

but they do not give a mathematical justification.

REFERENCE
Buckley and James, _Biometrika_ (1979).

NOTES

(i) The Buckley-James method is a nonparametric analogue of a normal theory technique due to Schmee and Hahn.
Define

$$W_i = \frac{T_i - \alpha - \beta x_i}{\sigma} \, .$$

If F is normal, then

$$E(T_i | T_i > y_i) = E\left(\sigma W_i + \alpha + \beta x_i \,\Big|\, W_i > \frac{y_i - \alpha - \beta x_i}{\sigma}\right) ,$$

$$= \alpha + \beta x_i + \frac{\sigma \int_{(y_i - \alpha - \beta x_i)/\sigma}^{\infty} w \, \phi(w) \, dw}{1 - \Phi\left(\frac{y_i - \alpha - \beta x_i}{\sigma}\right)} ,$$

$$= \alpha + \beta x_i + \frac{\sigma \, \phi\left(\frac{y_i - \alpha - \beta x_i}{\sigma}\right)}{1 - \Phi\left(\frac{y_i - \alpha - \beta x_i}{\sigma}\right)} ,$$

where ϕ and Φ are the standard normal density and distribution function, respectively. Schmee and Hahn use the estimate

$$\hat{E}(T_i | T_i > y_i) = \hat{\alpha} + \hat{\beta}x_i + \frac{\hat{\sigma} \; \phi\left(\dfrac{y_i - \hat{\alpha} - \hat{\beta}x_i}{\hat{\sigma}}\right)}{1 - \Phi\left(\dfrac{y_i - \hat{\alpha} - \hat{\beta}x_i}{\hat{\sigma}}\right)}$$

in place of (16).

REFERENCE
Schmee and Hahn, Technometrics (1979).

(ii) Both the parametric and nonparametric methods are analogous to the EM algorithm in maximum likelihood theory.

REFERENCE
Dempster, Laird, and Rubin, JRSS B (1977).

Koul-Susarla-Van Ryzin estimator. If we define

$$Y_i^* = \frac{\delta_i Y_i}{1 - G(Y_i)} \;,$$

then

$$E(Y_i^*) = \int_0^\infty \frac{u}{1 - G(u)} \, (1 - G(u)) \; dF_i(u) \;,$$

$$= \int_0^\infty u \; dF_i(u) \;,$$

$$= \alpha + \beta x_i \;.$$

Therefore, if we could observe $y_1^*, \ldots, y_n^*$, we could estimate α and β by (15) in the usual way.

Unfortunately, we cannot observe all of $y_1^*, \ldots, y_n^*$, but we can substitute estimates by replacing G with a PL estimator where the roles of the survival and censoring random variables are reversed. Alternatively, Koul, Susarla, and Van Ryzin propose to use a Bayesian estimator of G.

The big advantage with this method is that no iteration is required. Also, it is based on the assumption of a common censoring distribution rather than shifted censoring distributions as for the Miller estimator. However, the $\hat{y}_i^*$ are somewhat peculiar in that they are either zero or inflated values of the y_i. The behavior of this estimator has not been evaluated at this point.

REFERENCE
Koul, Susarla, and Van Ryzin, unpublished manuscript (1979).

Example. Stanford Heart Transplant Data. We compare the procedures of Cox, Miller, and Buckley-James when applied to the Stanford heart transplant data displayed in Table 5. In the first regression (Figure 5) the dependent variable is $\log_{10}$ survival time, where the survival time is the time until death due to rejection, and the covariate is the mismatch score. In the second regression (Figure 6) the dependent variable is $\log_{10}$ survival time, where the survival time here is the time until death, regardless of whether due to the rejection of the donor heart or other causes, and the covariate is age. In the one case where the survival time is recorded as zero this is changed to a 1 for taking logs. The comparisons of the three procedures are presented in Tables 6 and 7.

TABLE 5. STANFORD HEART TRANSPLANT DATA

Patient No.	Survival Time	Dead = 1 Alive = 0	Rejected = 1 Nonrejected = 0	Mismatch T5 Score	Age at Tx
3	15	1	0	1.11	54.3
4	3	1	0	1.66	40.4
7	624	1	1	1.32	51.0
10	46	1	1	0.61	42.5
11	127	1	0	0.36	48.0
13	64	1	1	1.89	54.6
14	1350	1	1	0.87	54.1
16	280	1	1	1.12	49.5
18	23	1	0	2.05	56.9
20	10	1	1	2.76	55.3
21	1024	1	1	1.13	43.4
22	39	1	1	1.38	42.8
23	730	1	1	0.96	58.4
24	136	1	1	1.62	52.0
25	1775	0	0	1.06	33.3
28	1	1	0	0.47	54.2
30	836	1	1	1.58	45.0
32	60	1	1	0.69	64.5
33	1536	0	0	0.91	49.0
34	1549	0	0	0.38	40.6

TABLE 5 (Continued)

Patient No.	Survival Time	Dead = 1 Alive = 0	Rejected = 1 Nonrejected = 0	Mismatch T5 Score	Age at Tx
36	54	1	1	2.09	49.0
37	47	1	1	0.87	61.5
38	0	1	0	0.87	41.5
39	51	1			50.5
40	1367	0	0	0.75	48.6
41	1264	0	0	0.98	45.5
45	44	1	0	0.0	36.2
46	994	1	1	0.81	48.6
47	51	1	1	1.38	47.2
49	1106	0	0	1.35	36.8
50	897	1			46.1
51	253	1	1	1.08	48.8
53	147	1			47.5
55	51	1	1	1.51	52.5
56	875	0	0	0.98	38.9
58	322	1	1	1.82	48.1
59	838	0	0	0.19	41.6
60	65	1	1	0.66	49.1
63	815	0	0	1.93	32.7
64	551	1	0	0.12	48.9

TABLE 5 (Continued)

Patient No.	Survival Time	Dead = 1 Alive = 0	Rejected = 1 Nonrejected = 0	Mismatch T5 Score	Age at Tx
65	66	1	1	1.12	51.3
67	228	1	0	1.02	19.7
68	65	1	1	1.68	45.2
69	660	0	0	1.20	48.0
70	25	1	1	1.68	53.0
71	589	0	0	0.97	47.5
72	592	0	0	1.46	26.7
73	63	1	1	2.16	56.4
74	12	1	0	0.61	29.2
76	499	0	0	1.70	52.2
78	305	0	0	0.81	49.3
79	29	1	1	1.08	54.0
80	456	0	0	1.41	46.5
81	439	0	0	1.94	52.9
83	48	1	0	3.05	53.4
84	297	1	1	0.60	42.8
86	389	0	0	1.44	48.9
87	50	1	1	2.25	46.4
88	339	0	0	0.68	54.4
89	68	1	1	1.33	51.4

TABLE 5 (Continued)

Patient No.	Survival Time	Dead = 1 Alive = 0	Rejected = 1 Nonrejected = 0	Mismatch T5 Score	Age at Tx
90	26	1	0	0.82	52.5
92	30	0	0	0.16	45.8
93	237	0	0	0.33	47.8
94	161	1	1	1.20	43.8
95	14	1			40.3
96	167	0	0	0.46	26.7
97	110	0	0	1.78	23.7
98	13	0	0	0.77	28.9
100	1	0	0	0.67	35.2

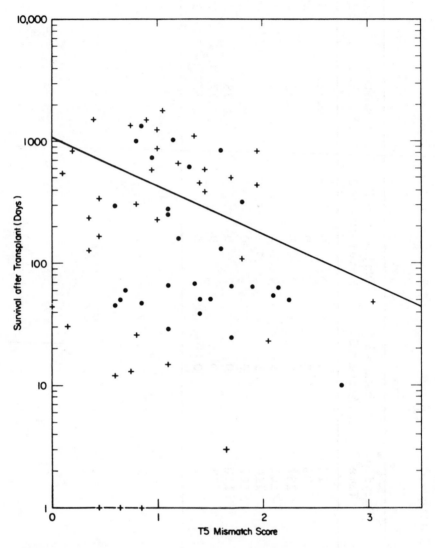

Figure 5. Survival versus T5 mismatch score. + = alive or nonrejection death, · = rejection death; —— Kaplan-Meier least squares line.

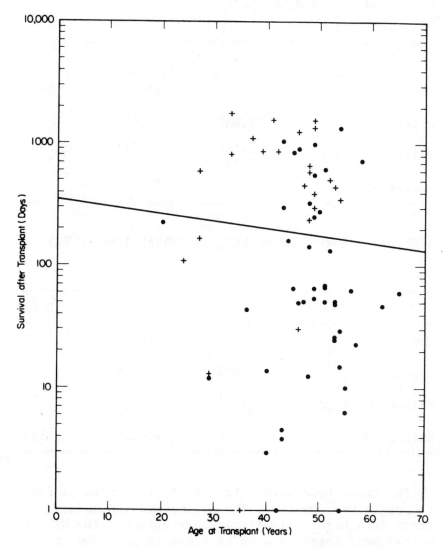

Figure 6. Survival versus age. + = alive, · = dead;
—— Kaplan-Meier least squares line.

161

TABLE 6. REGRESSION OF LOG_{10} SURVIVAL TIME UNTIL
REJECTION ON MISMATCH SCORE

	$\hat{\alpha}$	$\hat{\beta}$	$\hat{SD}(\hat{\beta})$
Cox		1.076	.368
Miller	3.036	-.394	
Miller modified	3.120	-.452	.236
	3.145	-.471	.234
Buckley-James			

TABLE 7. REGRESSION OF LOG_{10} SURVIVAL TIME ON AGE

	$\hat{\alpha}$	$\hat{\beta}$	$\hat{SD}(\hat{\beta})$
Cox		.0575	.0233
Miller	2.537	-.0058	
Miller modified	2.111	.0036	.0166
	2.171	.0024	.0163
Buckley-James	3.582	-.0278	.0149

   The three procedures give conflicting results for
the regression on age. The Cox method indicates
there is a highly significant age effect. The Miller
method says there is no effect due to age, and the
Buckley-James approach gives borderline significance
to age. The Miller estimators may be thrown off by
the censoring pattern in this case.

Since there is also disagreement about the degree of significance for the mismatch score effect, further work should be done to see which model (accelerated time or proportional hazards) is more appropriate for these data.

REFERENCES
Miller, Biometrika (1976).
Buckley and James, Biometrika (1979).

SEVEN

GOODNESS OF FIT

1. GRAPHICAL METHODS

The human eye can distinguish well between a straight line and a curve so the following basic principle should guide the method of plotting.

> Basic Principle. Select the scales of the co-ordinate axes so that if the model holds, a plot of the data resembles a straight line, and if the model fails, a plot resembles a curve.

There are two types of plots one can make, namely, survival plots and hazard plots. The two are closely related, and in each case the choice is one of convenience.

(i) Survival plots
Plot either

$$\hat{S}(y_{u(i)}) \text{ against } y_{u(i)} \text{ ,}$$

or

$\hat{S}(t)$ against   t .

This is a special case of   Q - Q   plots or probability plots.

REFERENCE
Wilk and Gnanadesikan, Biometrika (1968).

(ii) Hazard Plots
     Plot either

$\hat{\Lambda}(y_{u(i)})$ against $y_{u(i)}$,

or

$\hat{\Lambda}(t)$ against   t ,

using (see Chapter 3, Section 3) either the
Nelson formula

$$\hat{\Lambda}_2(t) = \sum_{y_{(i)} \leq t} \frac{\delta_{(i)}}{n - i + 1} ,$$

or

$$\hat{\Lambda}_1(t) = -\log \hat{S}(t) .$$

REFERENCES
Nelson, J. Qual. Tech. (1969).
_____, Technometrics (1972).

1.1. One Sample

Plots for several distributions are depicted on
the next page.

(i) <u>Exponential</u>

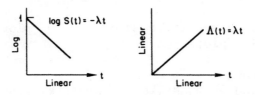

(ii) <u>Weibull</u>

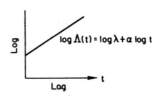

(iii) <u>Lognormal</u>

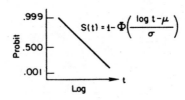

(iv) <u>Gamma and others</u>
   Without the use of special graph paper, compute and plot quantiles based on parametric assumptions against quantiles based on the PL estimator.

REFERENCE
Wilk, Gnanadesikan, and Huyett, <u>Technometrics</u> (1962), for the gamma distribution without censoring.

1.2. Two to K Samples

For parametric models, repeat (i) through (iv) on each sample.

Suppose we want to check the validity of the Cox proportional hazards model. Under the model,

$$S_i(t) = S_j(t)^{\gamma_{ij}}$$

for some γ_{ij} , so

$$\log S_i(t) = \gamma_{ij} \log S_j(t) ,$$

or

$$\frac{\log S_i(t)}{\log S_j(t)} = \gamma_{ij} .$$

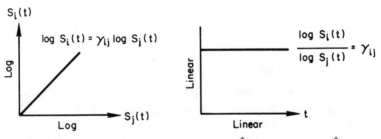

Compute individual PL estimates $\hat{S}_1(t), \ldots, \hat{S}_K(t)$, and form either of the above graphs.

To check the linear model, calculate the PL estimate for each of the K samples separately, and plot them, checking for shifts by translation.

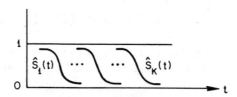

Example. DNCB Study. Hodgkin's disease patients
were sensitized and then continually exposed to the
chemical dinitrochlorobenzene (DNCB). The (+) population consists of those who react positively to ex-
posure to DNCB, and the (-) population consists of
those who do not react; patients can migrate between
populations. Survival time was taken to be time to
relapse.

Do the patients in the (+) population survive
longer than those in the (-) population? The Cox
proportional hazards model was used. The plot of
$\log \hat{S}^{(+)}(t)/\log \hat{S}^{(-)}(t)$ in Figure 7 shows us that
except for times t close to zero, the ratio of the
logs is reasonably constant, substantiating the validity of the model.

REFERENCE
Gong, Stanford Univ. Tech. Report No. 57 (1980).

1.3. Regression

Suppose we want to check the proportional hazards
model. In the case x is one-dimensional, we might
partition the x-axis into K intervals, compute a
separate PL estimator for each interval, and apply K-
sample procedures. If x is multidimensional, we
might try to partition the $\underset{\sim}{x}$-space into K regions.
However, grouping the data requires that the number
of observations be large and the required number of
observations grows rapidly with the dimensionality of
$\underset{\sim}{x}$.

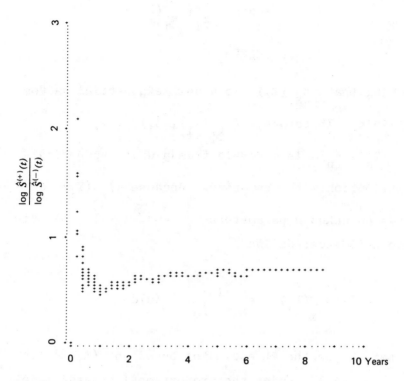

Figure 7. Computer graph of $\log \hat{S}^{(+)}(t)/\log \hat{S}^{(-)}(t)$ in DNCB study.

As an alternative to grouping, define

$$\Lambda_{\underset{\sim}{x}_i}(T_i) = e^{\underset{\sim}{\beta}' \underset{\sim}{x}_i} \int_0^{T_i} \lambda_0(u)\,du \ .$$

Then, under the proportional hazards model,

$$P\{\Lambda_{\underset{\sim}{x}_i}(T_i) > t\} = P\{T_i > \Lambda_{\underset{\sim}{x}_i}^{-1}(t)\} \ ,$$

$$= \exp\{-\Lambda_{\underset{\sim}{x}_i} (\Lambda_{\underset{\sim}{x}_i}^{-1}(t))\} \ ,$$

$$= e^{-t} \ ,$$

showing that $\Lambda_{\underset{\sim}{x}_i}(T_i)$ is a unit exponential random variable. Therefore, $(\Lambda_{\underset{\sim}{x}_1}(Y_1), \delta_1), \ \ldots,$ $(\Lambda_{\underset{\sim}{x}_n}(Y_n), \delta_n)$ is a sample from a unit exponential distribution with censoring. Because $\Lambda_{\underset{\sim}{x}_i}(Y_i)$ depends on unknown parameters $\underset{\sim}{\beta}$ and $\lambda_0(t)$, substitute estimates; define

$$\hat{\Lambda}_i = \hat{\Lambda}_{\underset{\sim}{x}_i}(Y_i) = e^{\hat{\underset{\sim}{\beta}}'\underset{\sim}{x}_i} \int_0^{Y_i} \hat{\lambda}_0(u)du \ .$$

Let $\hat{S}$ be the PL estimator based on $(\hat{\Lambda}_1, \delta_1),$ $\ldots, (\hat{\Lambda}_n, \delta_n)$. Under the proportional hazards model, $\log \hat{S}(t)$ should be approximately a linear function of t .

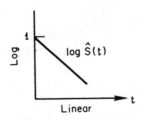

If the plot of $\log \hat{S}(t)$ against t is not linear, it may be difficult to guess what alternative model might be appropriate.

Under the proportional hazards model, $(\hat{\Lambda}_1, \delta_1)$,
..., $(\hat{\Lambda}_n, \delta_n)$ are (un)censored values of approximately iid random variables, so plotting $\hat{\Lambda}_i(t)$ a-
gainst a particular covariate x_{ij} or against $\hat{\underset{\sim}{\beta}}'\underset{\sim}{x}_i$
should not reveal any systematic patterns. The esti-
mates $\hat{\Lambda}_1, ..., \hat{\Lambda}_n$ are called <u>generalized residuals</u>
in the sense of Cox and Snell.

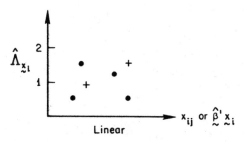

To check the linear model, if the number of observations is large, partition the $\underset{\sim}{x}$-region into K
subregions, and apply K-sample procedures. Alterna-
tively, plot the residuals $r_i = y_i - \hat{\underset{\sim}{\beta}}'\underset{\sim}{x}_i$ against a
particular covariate x_{ij} or against $\hat{\underset{\sim}{\beta}}'\underset{\sim}{x}_i$.

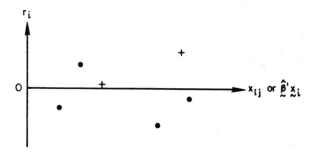

For both the proportional hazards and linear models the sensitivity of the residual plot to detect-
ing the correct model effect of a particular covariate x_{ij} may be enhanced by computing the residuals
without that covariate in the model.

REFERENCES
Cox and Snell, <u>JRSS B</u> (1968), discuss generalized
    residuals.
Crowley and Hu, <u>JASA</u> (1977), plot generalized re-
    siduals for the Stanford heart transplant data.
Kay, <u>Appl. Stat. (JRSS C)</u> (1977), discusses plot-
    ting generalized residuals.

2. TESTS

2.1. One Sample

We want to test

$$H_0 : F = F_0 , \quad F_0 \text{ specified.}$$

(i) <u>Generalized Kolmogorov (-Smirnov) test</u>
    Accept H_0 whenever

$$\sqrt{n}\,|\hat{F}(t) - F_0(t)| \le \hat{C}_n(t) \quad \text{for all} \quad t \ge 0 \; ,$$

where $\hat{F}(t)$ is the PL estimator and $\hat{C}_n(t)$
can be computed from tables. This test can be
used to construct simultaneous confidence bands
for $F_0(t)$:

$$P\left\{\hat{F}(t) - \frac{\hat{C}_n(t)}{\sqrt{n}} \leq F_0(t) \leq \hat{F}(t) + \frac{\hat{C}_n(t)}{\sqrt{n}}\right.$$

$$\left. \text{for all } t \geq 0 \right\} = 1 - \alpha .$$

REFERENCES
Barr and Davidson, Technometrics (1973), and
Koziol and Byar, Technometrics (1975), and
Dufour and Maag, Technometrics (1978), consider
    Type I and Type II censoring.
Gillespie and Fisher, Ann. Stat. (1979), and
Hall and Wellner, Biometrika (1980), consider
    the PL estimator and random censoring.

(ii) Generalized Cramér-von Mises test
    After performing a probability integral
transformation so that $F_0(t) = t$, the uniform

distribution function, the generalized Cramér-
von Mises test uses the statistic

$$n \int_0^1 (\hat{F}(t) - t)^2 dt ,$$

where $\hat{F}$ is the PL estimator.

REFERENCES
Koziol and Green, Biometrika (1976), consider
    the PL estimator and random censoring.
Pettit and Stephens, Biometrika (1976), consi-
    der Type I and Type II censoring. Pettit
    specializes to the normal and exponential
    distributions in
Pettit, Biometrika (1976), and
_____ , Biometrika (1977), respectively.

(iii) <u>Mantel-Haenszel type test</u>

REFERENCE
Hyde, <u>Biometrika</u> (1977).

(iv) <u>Limit of Efron's test</u>

REFERENCE
Hollander and Proschan, <u>Biometrics</u> (1979).

(v) <u>Parametric families</u>
Suppose we want to test

$$H_0 : F = F_{\underset{\sim}{\theta}} \, , \quad \underset{\sim}{\theta} \in \underset{\sim}{\Theta} \, .$$

The usual approach computes an estimate $\hat{\underset{\sim}{\theta}}$ and
checks if $\hat{F}$ is sufficiently close to $F_{\hat{\underset{\sim}{\theta}}}$.

REFERENCE
Mihalko and Moore, <u>Ann. Stat.</u> (1980), consider
χ^2-tests for Type II censoring with estimates that are asymptotically equivalent to
linear combinations of order statistics.

If $\underset{\sim}{\Theta}_0 \subset \underset{\sim}{\Theta}$, and we want to test

$$H_0 : \underset{\sim}{\theta} \in \underset{\sim}{\Theta}_0 \, ,$$

then a likelihood ratio test is appropriate.

REFERENCE
Turnbull and Weiss, <u>Biometrics</u> (1978), consider
likelihood ratio tests for discrete or
grouped data.

2.2. Regression

(i) <u>Parametric families</u>

Imbed the model in a larger model (e.g., a model that includes quadratic or cubic effects or interactions), and test to see if the smaller model holds. In effect, we are testing
$$H_0 : \underset{\sim}{\theta} \in \underset{\sim 0}{\Theta} \subset \underset{\sim}{\Theta}.$$

(ii) <u>χ^2-tests</u>

REFERENCES

Schoenfeld, <u>Biometrika</u> (1980), considers proportional hazards models with regions in the time × covariate space.

Lamborn, Stanford Univ. Tech. Report No. 21 (1969), looks at χ^2-tests for exponential regression.

EIGHT

MISCELLANEOUS TOPICS

1. BIVARIATE KAPLAN-MEIER ESTIMATOR

Let $T_i = (T_{i1}, T_{i2})$ be a pair of failure times. For example, they might be the times to failure of the left and right kidneys, or the times of cancer detection in the left and right breasts. Either or both times to failure may not be observable due to a one-dimensional random censoring variable C_i. The observable quantities are

$$Y_i = (Y_{i1}, Y_{i2}) = (T_{i1} \wedge C_i, T_{i2} \wedge C_i)$$

and the indicator vector

$$\delta = (\delta_{i1}, \delta_{i2}) = (I(T_{i1} \leq C_i), I(T_{i2} \leq C_i)) .$$

Muñoz has shown how to compute the two-dimensional generalization of the Kaplan-Meier estimator through the self-consistency and redistribute-to-the-right algorithms. In addition, he has established that

176

this estimator is the generalized maximum likelihood estimator and that it is a consistent estimator of the bivariate d.f. $F(t_1,t_2) = P\{T_{i1} \leq t_1, T_{i2} \leq t_2\}$.

Campbell considers the model with bivariate censoring times and treats the grouped data situation. Also, Korwar treats bivariate grouped data with both left and right censoring.

REFERENCES
Campbell, Purdue Univ. Mimeoseries #79-25 (1979), and
Korwar, unpublished manuscript (1980), treat bivariate grouped data with censoring.
Muñoz, Stanford Univ. Tech. Report No. 60 (1980), defines the two-dimensional KM estimator through algorithms and proves it is the GMLE.
_____, Stanford Univ. Tech. Report No. 61 (1980), proves consistency of the two-dimensional estimator.

2. COMPETING RISKS

Let $T_i = (T_{i1}, \ldots, T_{ip})$ be a p-dimensional vector of failure times. Each coordinate is the time to failure from a specific cause such as, for example, heart failure, cancer, kidney failure, and so on. The subject is observable only up to the time of the first failure. The failure times for all the other causes are censored by the failure of the system at the first failure time. The observable quantities are

$$T_i = \min\{T_{i1}, \ldots, T_{ip}\}$$

and

$$\delta_i = (\delta_{i1}, \ldots, \delta_{ip}) ,$$

$$= (I(T_{i1} \leq T_i), \ldots, I(T_{ip} \leq T_i)) \ .$$

The indicator vector $\underset{\sim}{\delta}$ denotes the specific cause of the failure.

The probability

$$P\{T_{ij} \leq t, \ \delta_{ij} = 1\}$$

is called the crude probability of dying from the cause j by time t. It is directly estimated by the observed proportion

$$\frac{1}{n} \sum_{i=1}^{n} I(T_i \leq t, \ \delta_{ij} = 1) \ .$$

The net probability is

$$P\{T_{ij} \leq t\} \ ,$$

and if the causes are independent, this can be consistently estimated by the PL method where all failure times other than from cause j are considered to be censoring times. Partial crude probabilities consider the probability of dying by time t from one of a subset of possible causes.

A fundamental result in the theory of competing risks is that on the basis of the sample T_i, $\underset{\sim}{\delta}_i$, $i = 1, \ldots, n$, it is impossible to tell whether

$$P\{T_{i1} \leq t_1, \ldots, T_{ip} \leq t_p\} = \prod_{j=1}^{p} P\{T_{ij} \leq t_j\}$$

or whether the failure times $T_{i1}, \ldots, T_{ip}$ are dependent. Different proofs of this result with varying conditions and degrees of rigor have appeared over the years. See the papers by Berman, Altshuler, Tsiatis, Peterson, and Langberg-Proschan-Quinzi.

REFERENCES

Chiang, Introduction to Stochastic Processes in Biostatistics (1968), discusses the relationships between crude, net, and partial crude probabilities in Chapter 11.

Moeschberger and David, Biometrics (1971), consider parametric likelihood methods.

Gail, Biometrics (1975), is a review article.

Prentice et al., Biometrics (1978), review competing risks from the hazard rate point of view.

Berman, Ann. Math. Stat. (1963),

Altshuler, Mathematical Biosciences (1970),

Tsiatis, Proc. Natl. Acad. Sci. (1975),

Peterson, Stanford Univ. Tech. Report No. 13 (1975),

_____, Proc. Natl. Acad. Sci. (1976), and

Langberg, Proschan, and Quinzi, Ann. Stat. (1981), examine the identifiability question.

3.   DEPENDENT CENSORING

Not much work has been done in the case where there is dependence between the failure times and the censoring times. Some of the work on dependent competing risks is relevant in this regard. Papers by Lagakos and Williams give some general discussion and results.

REFERENCES

Williams and Lagakos, Biometrika (1977).

Lagakos and Williams, Biometrika (1978).

Lagakos, Biometrics (1979).

4. JACKKNIFING AND BOOTSTRAPPING

Suppose that the parameter θ is a functional $T(F)$ of the d.f. F. In many instances θ is estimated by substituting the sample d.f. F_n for F in T, that is, $\hat{\theta} = T(F_n)$. For a sample $Y_1, \ldots, Y_n$, iid according to F, the <u>jackknifed estimate</u> $\tilde{\theta}$ of θ is defined as follows:

$$\tilde{\theta}_i = n\hat{\theta} - (n-1)\hat{\theta}_{-i}, \quad i = 1, \ldots, n,$$

$$\tilde{\theta} = \frac{1}{n} \sum_{i=1}^{n} \tilde{\theta}_i = n\hat{\theta} - \frac{(n-1)}{n} \sum_{i=1}^{n} \hat{\theta}_{-i},$$

where $\hat{\theta}_{-i} = T(F_{n-1,-i})$ is the estimate of θ with the i-th observation Y_i deleted from the sample.

When no censoring is present, it has been established that for sufficiently smooth T

$$\frac{\tilde{\theta} - \theta}{\sqrt{\frac{1}{n(n-1)} \Sigma_1^n (\tilde{\theta}_i - \tilde{\theta})^2}} \overset{a}{\sim} N(0, 1) . \qquad (18)$$

For confirmation of (18) in a variety of circumstances see Miller's 1974 review paper. Miller has also shown that (18) holds for randomly censored data with $\hat{\theta} = T(\hat{F})$ where $\hat{F}$ is the PL estimator.

The smoothness in T necessary for (18) to hold is connected with the smoothness of the influence function

$$IC(y;F) = \lim_{\varepsilon \to 0} \frac{T((1-\varepsilon)F + \varepsilon\delta_y) - T(F)}{\varepsilon},$$

where δ_y is the distribution function which puts mass one at y. For uncensored data the jackknife and influence function are related by

$$(n-1)(\hat{\theta} - \hat{\theta}_{-i})$$

$$= \left. \frac{T((1-\varepsilon)F + \varepsilon\delta_y) - T(F)}{\varepsilon} \right|_{\varepsilon=-1/(n-1), F=F_n, y=y_i}.$$

For censored data Reid has worked out the influence functions (i.e., partial derivatives with respect to F_u and F_c) for a function of the PL estimator.

Efron's <u>bootstrapping</u> is accomplished in the following manner. Let $Y_1^*, \ldots, Y_n^*$ be a sample with replacement from $y_1, \ldots, y_n$ when no censoring is present, and with censoring let $(Y_1^*, \delta_1^*), \ldots, (Y_n^*, \delta_n^*)$ be a sample with replacement from $(y_1, \delta_1), \ldots, (y_n, \delta_n)$. Then F_n^* and $\hat{F}^*$ are the <u>bootstrap sample</u> distribution function and PL estimator, respectively, and $\hat{\theta}^* = T(F_n^*)$ or $T(\hat{F}^*)$ depending on whether the data are uncensored or censored. This sampling procedure is repeated N times to produce $\hat{\theta}_1^*, \ldots, \hat{\theta}_N^*$. The empirical distribution of $\hat{\theta}_1^*, \ldots, \hat{\theta}_N^*$ is used to approximate the distribution of $\hat{\theta}$. More specifically, with pivotal quantities the empirical distribution of $\hat{\theta}^* - \hat{\theta}$ is used as an approximation to the distribution of $\hat{\theta} - \theta$.

REFERENCES

Miller, <u>Biometrika</u> (1974), reviews the jackknife
   for uncensored data problems.

_____, Stanford Univ. Tech. Report No. 14 (1975),
   establishes the validity of jackknifing the PL
   estimator.

Reid, <u>Ann. Stat.</u> (1981), derives the influence
   functions for the PL estimator.

Efron, <u>Ann. Stat.</u> (1979), introduces bootstrapping
   for uncensored data problems.

_____, Stanford Univ. Tech. Report No. 53 (1980),
   studies bootstrapping censored data in general
   and the median in particular.

NINE

PROBLEMS

Problem 1

Prove that the gamma distribution has IFR for $\alpha > 1$ and DFR for $\alpha < 1$.

Answer

$$\frac{1}{\lambda(t)} = \frac{\displaystyle\int_t^\infty x^{\alpha-1} e^{-\lambda x} dx}{t^{\alpha-1} e^{-\lambda t}} ,$$

$$= \int_t^\infty \left(\frac{x}{t}\right)^{\alpha-1} e^{-\lambda(x-t)} dx ,$$

$$= \int_0^\infty \left(1+\frac{u}{t}\right)^{\alpha-1} e^{-\lambda u} du \quad \text{(change of variable:}$$
$$u = x - t) .$$

If $\alpha > 1$,

$$\left(1 + \frac{u}{t}\right)^{\alpha-1}$$

is decreasing in t , so $\lambda(t)$ is increasing. For $\alpha < 1$ the integrand is increasing in t so $\lambda(t)$ is decreasing.

Problem 2

Derive the Fisher information for one observation from an exponential distribution with Type I censoring.

Answer

Let t_c be the fixed censoring time. The log of the likelihood is

$$\delta \log \lambda - \delta\lambda y - (1 - \delta)\lambda t_c .$$

Differentiating twice with respect to λ one gets

$$-\frac{\delta}{\lambda^2} ,$$

so the Fisher information is

$$I(\lambda) = \frac{1}{\lambda^2} E(\delta) = \frac{1}{\lambda^2} P\{T \le t_c\} = \frac{1}{\lambda^2} (1 - e^{-\lambda t_c}) .$$

Problem 3

Derive the sample information matrix for the Weibull distribution under random censoring.

Answer

From Example 2 in Chapter 2,

$$\frac{\partial}{\partial \gamma} \log L = \frac{n_u}{\gamma} - \sum_{i=1}^{n} y_i^{\alpha} ,$$

$$\frac{\partial}{\partial \alpha} \log L = \frac{n_u}{\alpha} + \sum_{u} \log t_i - \gamma \sum_{i=1}^{n} y_i^{\alpha} \log y_i .$$

The sample information matrix at (γ, α) is

$$-\begin{pmatrix} \dfrac{\partial^2}{\partial \gamma^2} \log L & \dfrac{\partial^2}{\partial \gamma\, \partial \alpha} \log L \\[4mm] & \dfrac{\partial^2}{\partial \alpha^2} \log L \end{pmatrix}$$

$$= \begin{pmatrix} \dfrac{1}{\gamma^2} n_u & \sum_{i=1}^{n} y_i^{\alpha}(\log y_i) \\[4mm] & \dfrac{1}{\alpha^2} n_u + \gamma \sum_{i=1}^{n} y_i^{\alpha}(\log y_i)^2 \end{pmatrix} .$$

Problem 4

From February 1972 to February 1975, 29 severe viral hepatitis patients satisfied the admission criteria for a 16 week study of the effects of steroid therapy at the Stanford, Veterans Administration, and Santa Clara Valley Hospitals and were randomized into either the steroid or control group. The survival times (in weeks) of the 14 patients in the steroid group were

1, 1, 1, 1+, 4+, 5, 7, 8, 10, 10+, 12+, 16+,

16+, 16+ .

Assume an exponential distribution $S(t) = \exp(-\lambda t)$.

(a) Estimate λ by maximum likelihood and construct an approximate 95% confidence interval.

(b) Estimate $S(16)$ and construct an approximate 95% confidence interval.

(c) Estimate the median survival time and construct an approximate 95% confidence interval.

REFERENCE
Gregory et al., <u>New England Journal of Medicine</u>
(1976).

<u>Answer</u>

(a) From Example 1 in Chapter 2, Section 2.1,

$$\hat{\lambda} = \frac{n_u}{\sum_{i=1}^{n} y_i} = \frac{7}{108} = .065 ,$$

and

$$\log \hat{\lambda} \overset{a}{\sim} N\left(\log \lambda, \frac{1}{n_u}\right) .$$

Then a 95% confidence interval for λ is given by

$$\left(\hat{\lambda} \exp\left(\frac{-Z_{.025}}{\sqrt{n_u}}\right), \ \hat{\lambda} \exp\left(\frac{Z_{.025}}{\sqrt{n_u}}\right)\right) = (.031, .136) .$$

(b) $\hat{S}(16) = \exp(-\hat{\lambda} \times 16) = .355.$
A 95% confidence interval for $S(16)$ is given by

$$(e^{-.136 \times 16}, \ e^{-.031 \times 16}) = (.113, .609) \ .$$

(c) $\hat{t}_{med} = (\log 2)/\hat{\lambda} = 10.69.$
A 95% confidence interval for the median is given by

$$\left(\frac{\log 2}{.136}, \ \frac{\log 2}{.031}\right) = (5.097, \ 22.36) \ .$$

Problem 5

For the severe viral hepatitis data in Problem 4 compute the Kaplan–Meier product–limit estimate of the survival function. Graph it and the survival function estimated under the exponential assumption on the same $\log \times$ linear graph paper. Do you think the assumption of an exponential distribution over the 16 week interval is justified?

Answer (see graph on next page)

$$\hat{S}(t) = \begin{cases} 1 & 0 \leq t < 1 \ , \\ 11/14 = .786 & 1 \leq t < 5 \ , \\ 11 \times 8/14 \times 9 = .7 & 5 \leq t < 7 \ , \\ 11 \times 7/14 \times 9 = .61 & 7 \leq t < 8 \ , \\ 11 \times 6/14 \times 9 = .524 & 8 \leq t < 10 \ , \\ 11 \times 5/14 \times 9 = .436 & 10 \leq t < 16 \ . \end{cases} \quad \text{for}$$

"Democratic" goodness of fit results: out of 21 student papers, 15 were in favor of the exponential, 5 were not, and 1 did not answer.

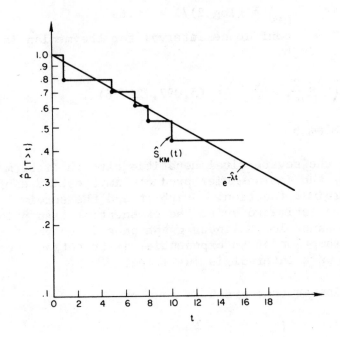

Problem 6

For the life table from Cutler and Ederer (Table 1) compute the approximate standard error of $\hat{S}(5)$.

Answer

Using Greenwood's formula

$$\hat{Var}(\hat{S}(5)) \cong (.44)^2 \left[\frac{47}{116.5(116.5 - 47)} \right.$$

$$+ \frac{5}{51.5(51.5 - 5)} + \frac{2}{30.5(30.5 - 2)}$$

$$+ \left. \frac{2}{16.5(16.5 - 2)} + \frac{0}{7(7 - 0)} \right],$$

$$= .003608 ,$$

so

$$\hat{SE}(\hat{S}(5)) \cong .06 .$$

Problem 7

For the Embury et al. length of remission AML data (Example in Chapter 3, Section 2) compute the approximate standard error of $\hat{S}(24)$ in the maintained group.

Answer

Using Greenwood's formula

$$\hat{Var}(\hat{S}(24)) = \left(\frac{6 \times 9}{11 \times 8} \right)^2 \left(\frac{1}{10 \times 11} + \frac{1}{9 \times 10} + \frac{1}{7 \times 8} \right.$$

$$+ \left. \frac{1}{6 \times 7} \right),$$

$$= .02329 ,$$

so

$$\hat{SE}(\hat{S}(24)) = .1526 .$$

Problem 8

Show in the proof that the PL estimator is the GMLE, the maximum of

$$
\prod_{i=1}^{n} P_i^{\delta(i)} \left(\sum_{j=i}^{n} P_j \right)^{1-\delta(i)}
$$

is attained for

$$
P_i = \frac{\delta(i)}{n - i + 1} \prod_{j=1}^{i-1} \left(1 - \frac{\delta(j)}{n - j + 1} \right).
$$

Answer

Let

$$
\lambda_i = \frac{P_i}{\sum_{j=i}^{n} P_j} , \quad i = 1, \ldots, n .
$$

Then, since $1 - \lambda_i = \sum_{j=i+1}^{n} P_j / \sum_{j=i}^{n} P_j$ and $\sum_{j=1}^{n} P_j = 1$, we have

$$
\sum_{j=i}^{n} P_j = \prod_{j=1}^{i-1} (1 - \lambda_j) ,
$$

and since $\lambda_n = 1$, we get

$$
\prod_{i=1}^{n} P_i^{\delta(i)} \left(\sum_{j=i}^{n} P_j \right)^{1-\delta(i)} = \prod_{i=1}^{n} \lambda_i^{\delta(i)} \prod_{j=1}^{i-1} (1 - \lambda_j) ,
$$

$$= \prod_{i=1}^{n-1} \lambda_i^{\delta_{(i)}} (1 - \lambda_i)^{n-i} .$$

It is well known from binomial sampling theory that each product is maximized by

$$\hat{\lambda}_i = \frac{\delta_{(i)}}{n - i + \delta_{(i)}} = \frac{\delta_{(i)}}{n - i + 1} .$$

Hence,

$$\hat{P}_i = \hat{\lambda}_i \left(\sum_{j=i}^{n} \hat{P}_j \right) = \hat{\lambda}_i \prod_{j=1}^{i-1} (1 - \hat{\lambda}_j)$$

$$= \frac{\delta_{(i)}}{n - i + 1} \prod_{j=1}^{i-1} \left(1 - \frac{\delta_{(i)}}{n - j + 1} \right) .$$

Problem 9

Prove that the redistribute-to-the-right algorithm gives the Kaplan-Meier product-limit estimator. Assume no ties.

Answer

There are two principal ways of proving this result.

(1) With the redistribute-to-the-right algorithm, all points $y_{(i)}$, censored or uncensored, initially have equal mass $1/n$. The algorithm moves from left to right through the order statistics. When it reaches $y_{(i)}-$, all the remaining points $y_{(i)}, y_{(i+1)}, \ldots, y_{(n)}$ have equal mass on them due to the way the algorithm operates. Suppose the total remaining

mass is $\tilde{S}(y_{(i)}-)$. By the equality of the
masses $y_{(i)}$ has $\tilde{S}(y_{(i)}-)/(n-i+1)$ assigned
to it, which it will keep if it is uncensored.
If it is censored, this mass is distributed to
the right.

Since the PL estimator $\hat{S}$ starts at 1 as
does $\tilde{S}$ and has jumps of sizes
$\hat{S}(y_{(i)}-)/(n-i+1)$ at the uncensored observations and zero at the censored observations,
the two estimators are identical.

(2) For the Kaplan-Meier estimator

$$\hat{\Delta}_{(i)} = \hat{S}(y_{(i)}-) - \hat{S}(y_{(i)}) ,$$

$$= \prod_{j=1}^{i-1} \left(\frac{n-j}{n-j+1}\right)^{\delta(j)} - \prod_{j=1}^{i} \left(\frac{n-j}{n-j+1}\right)^{\delta(j)} ,$$

$$= \prod_{j=1}^{i-1} \left(\frac{n-j}{n-j+1}\right)^{\delta(j)} \frac{\delta(i)}{n-i+1} ,$$

$$= \prod_{j=1}^{i-1} \left(\frac{n-j+1}{n-j}\right)^{-\delta(j)} \times \frac{1}{n} \times \frac{n}{n-1} \times \cdots$$

$$\times \frac{n-i+2}{1} \times \frac{\delta(i)}{n-i+1} ,$$

$$= \frac{\delta(i)}{n} \prod_{j=1}^{i-1} \left(\frac{n-j+1}{n-j}\right)^{1-\delta(j)} .$$

Let $j_1 < \cdots < j_i$ be the indices of the
censored observations which precede $y_{(i)}$. For
the redistribute-to-the-right algorithm the
mass assigned to $y_{(i)}$ if $\delta_{(i)} = 1$ is

$$\tilde{\Delta}_{(i)} = \frac{1}{n}\left(1 + \frac{1}{n-j_1}\right)\left(1 + \frac{1}{n-j_2}\right) \cdots \left(1 + \frac{1}{n-j_i}\right),$$

$$= \frac{1}{n} \prod_{j=1}^{i-1} \left(\frac{n-j+1}{n-j}\right)^{1-\delta}(j) ,$$

and if $\delta_{(i)} = 0$, $\tilde{\Delta}_{(i)} = 0$. This is identical to $\hat{\Delta}_{(i)}$, so the redistribute-to-the-right algorithm gives the PL estimator.

Problem 10

Given that for the PL estimator $\hat{S}(t)$

$$ACov(\hat{S}(t_1), \hat{S}(t_2)) = \frac{S(t_1) \, S(t_2)}{n}$$

$$\times \int_0^{t_1 \wedge t_2} \frac{dF_u(s)}{(1 - H(s))^2} ,$$

show that for $\hat{\mu} = \int_0^\infty \hat{S}(t)dt$

$$AVar(\hat{\mu}) = \frac{1}{n} \int_0^\infty \frac{1}{(1 - H(s))^2} \left(\int_s^\infty S(t)dt\right)^2 dF_u(s)$$

(where "AVar and ACov" denote the "asymptotic variance and covariance").

Answer

$$Var(\hat{\mu}) = E(\hat{\mu}^2) - (E(\hat{\mu}))^2 ,$$

$$= E\left(\int_0^\infty \int_0^\infty \hat{S}(t_1)\cdot\hat{S}(t_2)\, dt_1\, dt_2\right)$$

$$-\left(E\left(\int_0^\infty \hat{S}(t)dt\right)\right)^2 ,$$

$$= \int_0^\infty \int_0^\infty \text{Cov}(\hat{S}(t_1), \hat{S}(t_2))dt_1\, dt_2 ,$$

so

$$\text{AVar}(\hat{\mu}) = \frac{1}{n}\int_0^\infty \int_0^\infty S(t_1)\, S(t_2)$$

$$\times \int_0^{t_1 \wedge t_2} \frac{dF_u(s)}{(1-H(s))^2}\, dt_1\, dt_2 .$$

By interchanging integrals (applying Fubini's theorem)

$$\text{AVar}(\hat{\mu}) = \frac{1}{n}\int_0^\infty \frac{1}{(1-H(s))^2}\int_s^\infty S(t_1)dt_1$$

$$\times \int_s^\infty S(t_2)dt_2\, dF_u(s) ,$$

$$= \frac{1}{n}\int_0^\infty \frac{1}{(1-H(s))^2}\left(\int_s^\infty S(t)dt\right)^2 dF_u(s) .$$

Problem 11

For the Embury et al. AML data in the nonmaintained group, that is,

5, 5, 8, 8, 12, 16+, 23, 27, 30, 33, 43,

45 weeks ,

compute (a) $\hat{\mu}$ and (b) $\text{Var}(\hat{\mu})$.

Answer

(a) The Kaplan-Meier estimator $\hat{S}(t)$ is given in the following tables:

t $\in$	[0, 5)	[5, 8)	[8, 12)	[12, 23)	[23, 27)
$\hat{S}(t)=$	1	$\frac{10}{12}$	$\frac{8}{12}$	$\frac{7}{12}$	$\frac{7}{12} \times \frac{5}{6}$

t $\in$	[27, 30)	[30, 33)	[33, 43)	[43, 45)	[45, ∞)
$\hat{S}(t)=$	$\frac{7}{12} \times \frac{4}{6}$	$\frac{7}{12} \times \frac{3}{6}$	$\frac{7}{12} \times \frac{2}{6}$	$\frac{7}{12} \times \frac{1}{6}$	0

Then,

$$\hat{\mu} = \int_0^\infty \hat{S}(t)dt ,$$

$$= 1 \times 5 + \frac{10}{12} \times 3 + \frac{8}{12} \times 4 + \frac{7}{12} \times 11 + \frac{7}{12} \times \frac{5}{6} \times 4$$

$$+ \frac{7}{12} \times \frac{4}{6} \times 3 + \frac{7}{12} \times \frac{3}{6} \times 3 + \frac{7}{12} \times \frac{2}{6} \times 10$$

$$+ \frac{7}{12} \times \frac{1}{6} \times 2 ,$$

$$= 22.71 .$$

(b) $\text{Va\hat{r}}(\hat{\mu}) = \sum_{u} \left(\int_{y_{(i)}}^{\infty} \hat{S}(t)dt \right)^2 \frac{d_i}{n_i(n_i - d_i)}$,

$= (17.71)^2 \frac{2}{12 \times 10} + (15.21)^2 \frac{2}{10 \times 8}$

$+ (12.54)^2 \frac{1}{8 \times 7} + (6.125)^2 \frac{1}{6 \times 5}$

$+ (4.18)^2 \frac{1}{5 \times 4} + (3.01)^2 \frac{1}{4 \times 3}$

$+ (2.14)^2 \frac{1}{3 \times 2} + (.19)^2 \frac{1}{2 \times 1}$,

$= 17.47$.

Problem 12

For the Gregory et al. severe viral hepatitis data (see Problem 4) the steroid (I) and control (II) groups survival times (in weeks) are

I: 1, 1, 1, 1+, 4+, 5, 7, 8, 10, 10+, 12+,
    16+, 16+, 16+ ,

II: 1+, 2+, 3, 3, 3+, 5+, 5+, 16+(8) ,

with $m = 14$, $n = 15$. Compute

(a) the Gehan statistic,

(b) its permutation variance, and

(c) the normalized statistic and P-value.

REFERENCE
Gregory et al., New England Journal of Medicine
    (1976).

Answer

The U^* scores are computed in the following table:

Z	Group	# < Z	# > Z	U^*
1(3)	I	0	26	−26(3)
1+	I	3	0	3
1+	II	3	0	3
2+	II	3	0	3
3(2)	II	3	21	−18(2)
3+	II	5	0	5
4+	I	5	0	5
5	I	5	18	−13
5+(2)	II	6	0	6(2)
7	I	6	15	−9
8	I	7	14	−7
10	I	8	13	−5
10+	I	9	0	9
12+	I	9	0	9
16+(3)	I	9	0	9(3)
16+(8)	II	9	0	9(8)

(a) The Gehan statistic is

$$\sum_{II} U^* = 59 \ .$$

(b) Its permutation variance is

$$\frac{14 \times 15}{29 \times 28} \sum_{\text{I,II}} (\text{U}^*)^2 = 1086.72 \ .$$

(c) The normalized statistic is

$$\frac{59}{\sqrt{1086.72}} = 1.79 \ ,$$

which corresponds to a one-sided P-value of .0375.

Problem 13

For the Gregory et al. severe viral hepatitis data (see Problem 12) compute

(a) the Mantel-Haenszel statistic and its associated P-value, and

(b) the Tarone-Ware version of the Gehan statistic
and its associated P-value.

Answer

The computations in the table on the next page have been set up as in Table 4 of Chapter 4, Section 2.

(a)

$$\text{MH} = \frac{\text{sum of } a - E_0(A) \text{ column}}{\sqrt{\text{sum of } \left(\frac{m_1 (n-m_1)}{n-1} \text{ col.} \times \frac{n_1}{n}\left(1 - \frac{n_1}{n}\right) \text{ col.} \right)}} \ ,$$

$$= \frac{2.814}{\sqrt{2.1578}} = 1.916 \ ,$$

so the one-sided P-value is .027.

z	n	m_1	n_1	a	$E_0(A)$	$a - E_0(A)$	$n(a - E_0(A))$	$\dfrac{m_1(n-m_1)}{n-1}$	$\dfrac{n_1}{n}(1 - \dfrac{n_1}{n})$	$n_1(n - n_1)$
1	29	3	14	3	1.448	1.552	45	2.786	.2497	210
3	23	2	10	0	.869	-.869	-20	1.909	.2457	130
5	19	1	9	1	.474	.526	10	1	.2493	90
7	16	1	8	1	.500	.500	-8	1	.2500	64
8	15	1	7	1	.467	.533	8	1	.2489	56
10	14	1	6	1	.428	.572	8	1	.2449	48

(b) Let U_{TW} denote the Tarone–Ware version of the Gehan statistic.

$$U_{TW} = \frac{\text{sum of } n(a - E_0(A)) \text{ column}}{\sqrt{\text{sum of } \left(\frac{m_1(n-m_1)}{n-1} \text{ col.} \times n_1(n-n_1) \text{ col.} \right)}},$$

$$= \frac{59}{\sqrt{1091.23}} = 1.786 ,$$

so the one-sided P-value is .037.

Problem 14

As a prototype problem, consider the following five points from the Stanford heart transplant data (Example in Chapter 6, Section 2).

Mismatch Score (X)	Survival Time (Y)
2.09	54
.36	127+
.60	297
1.44	389+
.91	1536+

(a) Test the hypothesis $H_0 : \beta = 1$ in the proportional hazards model by calculating the P-value associated with the Cox statistic

$$\frac{\left(\frac{\partial}{\partial \beta} \log L_c(1) \right)^2}{- \frac{\partial^2}{\partial \beta^2} \log L_c(1)} .$$

(b) Compute the Tsiatis/Link estimate of $S(t; x)$
for $x = 1.5$ and $0 \le t \le 297$ using $\beta = 1$.

Answer

(a) Let

$$i : \quad 1 \quad 2 \quad 3 \quad 4 \quad 5$$
$$x_i : 2.09 \quad .36 \quad .60 \quad 1.44 \quad .91 .$$

From the expressions in Chapter 6, Section 1,

$$\frac{\partial}{\partial \beta} \log L_c(1) = x_1 + x_3 - \frac{\sum_{j=1}^{5} x_j e^{x_j}}{\sum_{j=1}^{5} e^{x_j}}$$

$$- \frac{\sum_{j=3}^{5} x_j e^{x_j}}{\sum_{j=3}^{5} e^{x_j}} ,$$

$$= .0965 ,$$

and

$$-\frac{\partial^2}{\partial \beta^2} \log L_c(1) = \frac{\sum_{j=1}^{5} x_j^2 e^{x_j}}{\sum_{j=1}^{5} e^{x_j}} - \left(\frac{\sum_{j=1}^{5} x_j e^{x_j}}{\sum_{j=1}^{5} e^{x_j}} \right)^2$$

$$+ \frac{\displaystyle\sum_{j=3}^{5} x_j^2\, e^{x_j}}{\displaystyle\sum_{j=3}^{5} e^{x_j}} - \left(\frac{\displaystyle\sum_{j=3}^{5} x_j\, e^{x_j}}{\displaystyle\sum_{j=3}^{5} e^{x_j}} \right)^2 ,$$

$$= .5110 .$$

Thus,

$$\frac{\left(\frac{\partial}{\partial\beta} \log L_c(1) \right)^2}{- \frac{\partial^2}{\partial\beta^2} \log L_c(1)} = .018 ,$$

and the P-value from a χ_1^2 table is approximately .9.

(b) The Tsiatis estimate (see Chapter 6, Section 1) is

$$\hat{\Lambda}_{0,T}(t) = \begin{cases} 1 & \text{for } 0 \le t < 54 , \\[2em] \dfrac{1}{\displaystyle\sum_{j=1}^{5} e^{x_j}} = .0554 & \text{for } 54 \le t < 297 , \\[2em] \dfrac{1}{\displaystyle\sum_{j=1}^{5} e^{x_j}} + \dfrac{1}{\displaystyle\sum_{j=3}^{5} e^{x_j}} \\[1em] \qquad\qquad = .1727 & \text{for } t = 297 , \end{cases}$$

so

$$\hat{S}_T(t; 1.5) = e^{-\hat{\Lambda}_{0,T}(t)e^{1.5}},$$

$$= \begin{cases} 1 & \text{for } 0 \leq t < 54 \text{ ,} \\ .78 & \text{for } 54 \leq t < 297 \text{ ,} \\ .46 & \text{for } t = 297 \text{ .} \end{cases}$$

The Link estimate (see Chapter 6, Section 1) is

$$\hat{\Lambda}_{0,L}(t) = \begin{cases} \dfrac{.0554}{54} \, t & \text{for } 0 \leq t < 54 \text{ ,} \\ \dfrac{.1727 - .0554}{297 - 54} (t - 54) + .0554 & \\ & \text{for } 54 \leq t \leq 297 \text{ ,} \end{cases}$$

so

$$\hat{S}_L(t; 1.5) = e^{-\hat{\Lambda}_{0,L}(t)e^{1.5}}$$

$$= \begin{cases} e^{-.0046t} & \text{for } 0 \leq t < 54 \text{ ,} \\ .877 \, e^{-.0022t} & \text{for } 54 \leq t \leq 297 \text{ .} \end{cases}$$

Problem 15

For the Embury et al. AML data (Example in Chapter 3, Section 2)

Maintained group

9, 13, 13+, 18, 23, 28+, 31, 34, 45+, 48, 161+

Nonmaintained group

5, 5, 8, 8, 12, 16+, 23, 27, 30, 33, 43, 45

compare the two groups by

(a) the Gehan statistic and the Mantel permutation variance,

(b) the Mantel-Haenszel statistic, and

(c) the Tarone-Ware version of the Gehan statistic.

In each case obtain the normalized statistic and its associated two-sided P-value.

Answer

(a) For the computation of the scores needed to perform the Gehan statistic the table on the next page is useful. The Gehan statistic is

$$\sum_{NM} U^* = -50 \; ,$$

and the Mantel permutation variance is

$$\frac{11 \times 12}{23 \times 22} \sum_{M,NM} (U^*)^2 = 912 \; ,$$

so the normalized statistic is

$$\frac{-50}{\sqrt{912}} = -1.656 \; ,$$

which corresponds to a two-sided P-value of .098.

Z	Group	# < Z	# > Z	U*
5(2)	NM	0	21	−21(2)
8(2)	NM	2	19	−17(2)
9	M	4	18	−14
12	NM	5	17	−12
13	M	6	16	−10
13+	M	7	0	7
16+	NM	7	0	7
18	M	7	13	−6
23	M	8	11	−3
23	NM	8	11	−3
27	NM	10	10	0
28+	M	11	0	11
30	NM	11	8	3
31	M	12	7	5
33	NM	13	6	7
34	M	14	5	9
43	NM	15	4	11
45	NM	16	3	13
45+	M	17	0	17
48	M	17	1	16
161+	M	18	0	18

(b) and (c)   The computations in the table on
the next page are organized like those in the
Answer to Problem 13.   The Mantel-Haenszel normalized statistic is (see Answer to Problem 13)

z	n	m_1	n_1	a	$E_0(A)$	$a - E_0(A)$	$n(a - E_0(A))$	$\frac{m_1(n-m_1)}{n-1}$	$\frac{n_1}{n}(1 - \frac{n_1}{n})$	$n_1(n - n_1)$
5	23	2	11	0	.9565	-.9565	-22	1.909	.2495	132
8	21	2	11	0	1.048	-1.048	-22	1.9	.2494	110
9	19	1	11	1	.579	.421	8	1	.2437	88
12	18	1	10	0	.555	-.555	-10	1	.2469	80
13	17	1	10	1	.588	.412	7	1	.2422	70
18	14	1	8	1	.571	.429	6	1	.2449	48
23	13	2	7	1	1.077	-.077	-1	1.83	.2485	42
27	11	1	6	0	.545	-.545	-6	1	.2479	30
30	9	1	5	0	.555	-.555	-5	1	.2469	20
31	8	1	5	1	.625	.375	3	1	.2344	15
33	7	1	4	0	.571	-.571	-4	1	.2449	12
34	6	1	4	1	.667	.333	2	1	.2222	8
43	5	1	3	0	.6	-.6	-3	1	.2400	6
45	4	1	3	0	.75	-.75	-3	1	.1875	3
48	2	1	2	1	1.0	0	0	1	0	0

$$\frac{-3.69}{\sqrt{4.0072}} = -1.84 \ ,$$

so the two-sided P-value is .066.

The Tarone-Ware version of the Gehan statistic is (see Answer to Problem 13)

$$\frac{-50}{\sqrt{917.97}} = -1.65 \ ,$$

so the two-sided P-value is .099.

Problem 16

Prove that the numerator in the Tarone-Ware version of the Gehan statistic (i.e., $\Sigma \ n_i(a_i - E_0(A_i))$) equals the numerator of the Gehan statistic as defined by Gehan (i.e., $U = \Sigma\Sigma U_{ij}$), except possibly for a factor of -1. Allow for ties.

Answer

From Chapter 4, Section 1,

$$U = \sum_{k=1}^{m+n} U_k^* \ I(k \in I_1) \ ,$$

where

$$U_k^* = \sum_{\substack{\ell=1 \\ \ell \neq k}}^{m+n} U_{k\ell} \ ,$$

so if the observation with label k in sample 1 is uncensored, then U_k^* is equal to the

number of uncensored observations before it
minus the number of observations after it.

$$U_k^* = \sum_{j=1}^{k-1} m_{j1} - (n_k - m_{k1}) ,$$

$$= \sum_{j=1}^{k} m_{j1} - n_k \quad \text{(see Chapter 4, Section 2 for notation) .}$$

On the other hand, if the observation k is
censored, then U_k^* is the number of uncensored
observations before it.

$$U_k^* = \sum_{j=1}^{k} m_{j1} .$$

Therefore,

$$U = \sum_{\substack{k=1 \\ c}}^{m+n} \sum_{j=1}^{k} m_{j1} \, I(k \in I_1)$$

$$+ \sum_{\substack{k=1 \\ u}}^{m+n} \left(\sum_{j=1}^{k} m_{j1} - n_k \right) I(k \in I_1) ,$$

where c and u mean the sums are performed
over censored and uncensored observations,
respectively. Thus,

$$U = \sum_{k=1}^{m+n} \sum_{j=1}^{k} m_{j1} \, I(k \in I_1)$$

$$- \sum_{k=1}^{m+n} n_k \, I(k \in I_1, \; \delta_k = 1) \; ,$$

$$= \sum_{j=1}^{m+n} m_{j1} \sum_{k=j}^{m+n} I(k \in I_1) - \sum_{k=1}^{m+n} n_k \, a_k \; ,$$

$$= \sum_{j=1}^{m+n} (m_{j1} n_{j1} - n_j a_j) \; ,$$

$$= \sum_{u} (m_{j1} n_{j1} - n_j a_j) \; ,$$

$$= - \sum_{u} n_j (a_j - E_0(A_j)) \; ,$$

where the next to the last equality follows from the fact that m_j (hence a_j), which is the number of uncensored observations at z_j, is zero if z_j is a censored observation. (Recall the convention that ties between censored and uncensored observations are broken by considering the censored observations to be larger.)

Problem 17

Show that Mantel's permutation variance for the Gehan statistic, divided by $N^3 = (m+n)^3$, that is,

$$\frac{1}{N^3} \times \frac{mn}{(m+n)(m+n-1)} \sum_{i=1}^{m+n} (U_i^*)^2 \; ,$$

converges to

$$\lambda(1 - \lambda) \int_0^\infty (1 - H(t))^2 \, dH_u(t)$$

as $N \to \infty$, $m/N \to \lambda$ under the null hypothesis
$H_0^* : F_1 = F_2; \ G_1 = G_2$, where

$$H(t) = P\{Z \leq t\} = \int_0^t (1 - G(u)) \, dF(u)$$

$$+ \int_0^t (1 - F(u)) \, dG(u) \ ,$$

$$H_u(t) = P\{Z \leq t, \ \zeta = 1\} = \int_0^t (1 - G(u)) \, dF(u) \ ,$$

and F and G are continuous.

Answer

Let

$$\hat{H}(t) = \frac{1}{N} \sum_{i=1}^N I(Z_i \leq t) \ ,$$

$$\hat{H}_u(t) = \frac{1}{N} \sum_{i=1}^N I(Z_i \leq t, \ \zeta_i = 1) \ .$$

Then,

$$U_i^* = \begin{cases} \#(\text{uncensored obs.} < Z_i) - \#(\text{obs.} > Z_i) \\ \hspace{3cm} \text{if} \quad \zeta_i = 1 \ , \\ \#(\text{uncensored obs.} < Z_i) \quad \text{if} \quad \zeta_i = 0 \ , \end{cases}$$

$$= \begin{cases} N\,\hat{H}_u(Z_i-) - N(1-\hat{H}(Z_i)) \quad \text{if} \quad \zeta_i = 1 \ , \\ N\,\hat{H}_u(Z_i-) \hspace{2.5cm} \text{if} \quad \zeta_i = 0 \ , \end{cases}$$

$$= N[\hat{H}_u(Z_i-) - \zeta_i(1 - \hat{H}(Z_i))] \ .$$

Consequently,

$$\frac{1}{N^3} \sum_{i=1}^{N} (U_i^*)^2 = \frac{1}{N^3} \sum_{i=1}^{N} N^2[\hat{H}_u(Z_i-) - \zeta_i(1-\hat{H}(Z_i))]^2 \ ,$$

$$= \frac{1}{N} \sum_{i=1}^{N} \hat{H}_u^2(Z_i-)$$

$$- \frac{2}{N} \sum_{i=1}^{N} \zeta_i \, \hat{H}_u(Z_i-)(1 - \hat{H}(Z_i))$$

$$+ \frac{1}{N} \sum_{i=1}^{N} \zeta_i(1 - \hat{H}(Z_i))^2 \ ,$$

$$= \int_0^\infty \hat{H}_u^2(t-)\,d\hat{H}(t)$$

$$- 2 \int_0^\infty \hat{H}_u(t-)(1 - \hat{H}(t))\,d\hat{H}_u(t)$$

$$+ \int_0^\infty (1 - \hat{H}(t))^2 \, d\hat{H}_u(t) \ .$$

Since $\hat{H}(t) \overset{a.s.}{\to} H(t)$ and $\hat{H}_u(t) \overset{a.s.}{\to} H_u(t)$ uniformly in t as $N \to \infty$ by the Glivenko-Cantelli theorem, it follows that as $N \to \infty$

$$\frac{1}{N^3} \sum_{i=1}^{N} (U_i^*)^2 \overset{a.s.}{\to} \int_0^\infty H_u^2(t) dH(t)$$

$$- 2 \int_0^\infty H_u(t)(1 - H(t)) dH_u(t)$$

$$+ \int_0^\infty (1 - H(t))^2 dH_u(t) \ .$$

Integration by parts gives

$$2 \int_0^\infty H_u(t)(1 - H(t)) dH_u(t) = H_u^2(t)(1 - H(t)) \Big|_0^\infty$$

$$+ \int_0^\infty H_u^2(t) dH(t) \ ,$$

$$= \int_0^\infty H_u^2(t) dH(t) \ ,$$

so the first two terms in the above limiting expression cancel. This, together with

$$\frac{mn}{(m + n)(m + n - 1)} \to \lambda(1 - \lambda) \ ,$$

establishes the result.

REFERENCES

The numbers in brackets after the references are the numbers of the pages on which the references are cited.

Aalen, O. (1976). Nonparametric inference in connection with multiple decrement models. Scandinavian Journal of Statistics 3, 15-27. [69]

_____ (1978). Nonparametric inference for a family of counting processes. Annals of Statistics 6, 701-726. [69]

Abelson, R. P. and Tukey, J. W. (1963). Efficient utilization of non-numerical information in quantitative analysis: General theory and the case of simple order. Annals of Mathematical Statistics 34, 1347-1369. [112]

Altshuler, B. (1970). Theory for the measurement of competing risks in animal experiments. Mathematical Biosciences 6, 1-11. [179]

213

Bailey, K. R. (1979). The general maximum likelihood
    approach to the Cox regression model. Ph.D.
    dissertation, University of Chicago, Chicago,
    Illinois. [133]

Barlow, R. E. and Proschan, F. (1975). Statistical
    Theory of Reliability and Life Testing. Holt,
    Rinehart, and Winston, New York. [15]

Barr, D. R. and Davidson, T. (1973). A Kolmogorov-
    Smirnov test for censored samples. Technometrics
    15, 739-757. [173]

Basu, A. P. (1964). Estimates of reliability for
    some distributions useful in life testing.
    Technometrics 6, 215-219. [35]

Berkson, J. and Gage, R. P. (1950). Calculation of
    survival rates for cancer. Proceedings of the
    Staff Meetings of the Mayo Clinic 25, 270-286.
    [46]

Berman, S. M. (1963). Note on extreme values, com-
    peting risks and semi-Markov processes. Annals
    of Mathematical Statistics 34, 1104-1106. [179]

Billingsley, P. (1968). Convergence of Probability
    Measures. Wiley, New York. [65]

Breslow, N. (1970). A generalized Kruskal-Wallis
    test for comparing K samples subject to un-
    equal patterns of censorship. Biometrika 57,
    579-594. [109, 114]

_____ (1972). Discussion on Professor Cox's paper.
    Journal of the Royal Statistical Society, Series
    B 34, 216-217. [136]

_____ (1974). Covariance analysis of censored survival data. Biometrics 30, 89-99. [136, 139]

_____ and Crowley, J. (1974). A large sample study of the life table and product limit estimates under random censorship. Annals of Statistics 2, 437-453. [46, 65, 69]

Buckley, J. and James, I. (1979). Linear regression with censored data. Biometrika 66, 429-436. [153, 163]

Campbell, G. (1979). Nonparametric bivariate estimation with randomly censored data. Mimeoseries #79-25, Department of Statistics, Purdue University, West Lafayette, Indiana. [177]

Chiang, C. L. (1968). Introduction to Stochastic Processes in Biostatistics. Wiley, New York. [46, 179]

Cohen, A. C. (1965). Maximum likelihood estimation in the Weibull distribution based on complete and on censored samples. Technometrics 7, 579-588. [29]

Cox, D. R. (1972). Regression models and life-tables. Journal of the Royal Statistical Society, Series B 34, 187-202. [127, 139]

_____ (1975). Partial likelihood. Biometrika 62, 269-276. [132]

_____ and Snell, E. J. (1968). A general definition of residuals. Journal of the Royal Statistical Society, Series B 30, 248-275. [172]

Crowley, J. (1974). Asymptotic normality of a new nonparametric statistic for use in organ transplant studies. Journal of the American Statistical Association 69, 1006-1011. [103]

_____ and Hu, M. (1977). Covariance analysis of heart transplant survival data. _Journal of the American Statistical Association_ 72, 27-36. [141, 172]

Cutler, S. J. and Ederer, F. (1958). Maximum utilization of the life table method in analyzing survival. _Journal of Chronic Diseases_ 8, 699-712. [42, 46]

Dempster, A. P., Laird, N. M. and Rubin, D. B. (1977). Maximum likelihood from incomplete data via the EM algorithm. _Journal of the Royal Statistical Society, Series B_ 39, 1-22. [154]

Dufour, R. and Maag, U. R. (1978). Distribution results for modified Kolmogorov-Smirnov statistics for truncated or censored samples. _Technometrics_ 20, 29-32. [173]

Efron, B. (1967). The two sample problem with censored data. _Proceedings of the Fifth Berkeley Symposium on Mathematical Statistics and Probability, Vol. IV._ University of California Press, Berkeley, California. 831-853. [52, 57, 106]

_____ (1977). The efficiency of Cox's likelihood function for censored data. _Journal of the American Statistical Association_ 72, 557-565. [132]

_____ (1979). Bootstrap methods: Another look at the jackknife. _Annals of Statistics_ 7, 1-26. [182]

_____ (1980). Censored data and the bootstrap. Technical Report No. 53 (R01 GM21215), Division of Biostatistics, Stanford University, Stanford, California. [76, 182]

_____ and Hinkley, D. V. (1978). Assessing the accuracy of the maximum likelihood estimator: Observed versus expected Fisher information. Biometrika 65, 457-487. [21]

Elveback, L. (1958). Estimation of survivorship in chronic disease: The "actuarial" method. Journal of the American Statistical Association 53, 420-440. [46]

Embury, S. H., Elias, L., Heller, P. H., Hood, C. E., Greenberg, P. L. and Schrier, S. L. (1977). Remission maintenance therapy in acute myelogenous leukemia. Western Journal of Medicine 126, 267-272. [50]

Epstein, B. and Sobel, M. (1953). Life testing. Journal of the American Statistical Association 48, 486-502. [25]

Farewell, V. T. (1977). A model for a binary variable with time-censored observations. Biometrika 64, 43-46. [38]

Feigl, P. and Zelen, M. (1965). Estimation of exponential survival probabilities with concomitant information. Biometrics 21, 826-838. [36]

Ferguson, T. S. (1973). A Bayesian analysis of some nonparametric problems. Annals of Statistics 1, 209-230. [79]

_____ and Phadia, E. G. (1979). Bayesian nonparametric estimation based on censored data. Annals of Statistics 7, 163-186. [79]

Fleming, T. R. and Harrington, D. P. (1979). Nonparametric estimation of the survival distribution in censored data. Unpublished manuscript. [67]

Földes, A., Rejtö, L. and Winter, B. B. (1978).
    Strong consistency properties of nonparametric
    estimators for randomly censored data. Part II:
    Estimation of density and failure rate. Unpub-
    lished manuscript. [76]

Gail, M. (1975). A review and critique of some mo-
    dels used in competing risk analysis. Biometrics
    31, 209-222. [179]

Gaver, D. P., Jr. and Hoel, D. G. (1970). Comparison
    of certain small-sample Poisson probability es-
    timates. Technometrics 12, 835-850. [35]

Gehan, E. A. (1965). A generalized Wilcoxon test for
    comparing arbitrarily singly-censored samples.
    Biometrika 52, 203-223. [89]

Gilbert, J. P. (1962). Random censorship. Ph.D.
    dissertation, University of Chicago, Chicago,
    Illinois. [94]

Gillespie, M. J. and Fisher, L. (1979). Confidence
    bands for the Kaplan-Meier survival curve esti-
    mate. Annals of Statistics 7, 920-924. [173]

Glasser, M. (1967). Exponential survival with co-
    variance. Journal of the American Statistical
    Association 62, 561-568. [36]

Gong, G. (1980). Do Hodgkin's disease patients with
    DNCB sensitivity survive longer? Biostatistics
    Casebook, Vol. III, Technical Report No. 57
    (R01 GM21215), Division of Biostatistics, Stan-
    ford University, Stanford, California. [168]

Gregory, P. B., Knauer, C. M., Kempson, R. L. and
    Miller, R. (1976). Steroid therapy in severe
    viral hepatitis. New England Journal of Medi-
    cine 294, 687-690. [186, 196]

Gross, A. J. and Clark, V. A. (1975). Survival Distributions: Reliability Applications in the
Biomedical Sciences. Wiley, New York. [20]

Hall, W. J. and Wellner, J. A. (1980). Confidence
bands for a survival curve from censored data.
Biometrika 67, 133-143. [173]

Hollander, M. and Proschan, F. (1979). Testing to
determine the underlying distribution using randomly censored data. Biometrics 35, 393-401.
[174]

Hyde, J. (1977). Testing survival under right censoring and left truncation. Biometrika 64, 225-
230. [174]

_____ (1977). Life testing with incomplete observa-
tions. Technical Report No. 30 (R01 GM21215),
Division of Biostatistics, Stanford University,
Stanford, California. [94]

Johansen, S. (1978). The product limit estimator as
maximum likelihood estimator. Scandinavian
Journal of Statistics 5, 195-199. [59]

Johns, M. V., Jr. and Lieberman, G. J. (1966). An
exact asymptotically efficient confidence bound
for reliability in the case of the Weibull distribution. Technometrics 8, 135-175. [32]

Kalbfleisch, J. and Prentice, R. L. (1972). Discussion on Professor Cox's paper. Journal of
the Royal Statistical Society, Series B 34, 215-
216. [139]

_____ and _____ (1973). Marginal likelihoods based
on Cox's regression and life model. Biometrika
60, 267-278. [130]

_____ and _____ (1980). The Statistical Analysis of
Failure Time Data. Wiley, New York. [20, 127,
143, 146]

Kaplan, E. L. and Meier, P. (1958). Nonparametric
estimation from incomplete observations. Journal of the American Statistical Association 53,
457-481. [48, 59, 72]

Kay, R. (1977). Proportional hazard regression models and the analysis of censored survival data.
Applied Statistics (Journal of the Royal Statistical Society, Series C) 26, 227-237. [172]

Kiefer, J. and Wolfowitz, J. (1956). Consistency of
the maximum likelihood estimator in the presence
of infinitely many incidental parameters. Annals
of Mathematical Statistics 27, 887-906. [59]

Korwar, R. M. (1980). Nonparametric estimation of a
bivariate survivorship function with doubly censored data. Unpublished manuscript. [177]

Koul, H., Susarla, V. and Van Ryzin, J. (1979). Regression analysis with randomly right censored
data. Unpublished manuscript. [155]

Koziol, J. A. and Byar, D. P. (1975). Percentage
points of the asymptotic distributions of one
and two sample K-S statistics for truncated or
censored data. Technometrics 17, 507-510. [173]

_____ and Green, S. B. (1976). A Cramér-von Mises
statistic for randomly censored data. Biometrika 63, 465-474. [173]

Lagakos, S. W. (1979). General right censoring and
its impact on the analysis of survival data.
Biometrics 35, 139-156. [179]

_____ and Williams, J. S. (1978). Models for censored survival analysis: A cone class of variable-sum models. Biometrika 65, 181-189. [179]

Lamb, E. J. and Leurgans, S. (1979). Does adoption affect subsequent fertility? American Journal of Obstetrics and Gynecology 134, 138-144. [141]

Lamborn, K. (1969). On chi-squared goodness of fit tests for sampling from more than one population with possibly censored data. Technical Report No. 21 (T01 GM00025), Department of Statistics, Stanford University, Stanford, California. [175]

Langberg, N. A., Proschan, F. and Quinzi, A. J. (1981). Estimating dependent life lengths, with applications to the theory of competing risks. Annals of Statistics 9, 157-167. [179]

Latta, R. B. (1977). Generalized Wilcoxon statistics for the two-sample problem with censored data. Biometrika 64, 633-635. [146]

Leavitt, S. S. and Olshen, R. A. (1974). The insurance claims adjuster as patients' advocate: Quantitative impact. Report for Insurance Technology Company, Berkeley, California. [2]

Leiderman, P. H., Babu, D., Kagia, J., Kraemer, H. C. and Leiderman, G. F. (1973). African infant precocity and some social influences during the first year. Nature 242, 247-249. [7]

Leurgans, S. (1980). Does adoption affect fertility? A proportional hazards model. Biostatistics Casebook, Vol. III, Technical Report No. 57 (R01 GM21215), Division of Biostatistics, Stanford University, Stanford, California. [141]

Lininger, L., Gail, M. H., Green, S. B. and Byar, D.
    P. (1979). Comparison of four tests for equal-
    ity of survival curves in the presence of stra-
    tification and censoring. Biometrika 66, 419-
    428. [103]

Link, C. L. (1979). Confidence intervals for the
    survival function using Cox's proportional ha-
    zard model with covariates. Technical Report
    No. 45 (R01 GM21215), Division of Biostatistics,
    Stanford University, Stanford, California. [136]

Mantel, N. (1967). Ranking procedures for arbitrari-
    ly restricted observation. Biometrics 23, 65-
    78. [89]

_____ and Haenszel, W. (1959). Statistical aspects
    of the analysis of data from retrospective stu-
    dies of disease. Journal of the National Cancer
    Institute 22, 719-748. [103]

_____ and Myers, M. (1971). Problems of convergence
    of maximum likelihood iterative procedures in
    multiparameter situations. Journal of the
    American Statistical Association 66, 484-491.
    [36]

Marcuson, R. and Nordbrock, E. (1981). A K-sample
    generalization of the Gehan-Gilbert procedure
    for the analysis of arbitrarily censored survi-
    val data. Biometrische Zeitschrift/Biometrical
    Journal. [113]

Meier, P. (1975). Estimation of a distribution func-
    tion from incomplete observations. Perspectives
    in Probability and Statistics. Papers in Honour
    of M. S. Bartlett (Ed. J. Gani). Academic Press,
    New York. 67-82. [72]

Mihalko, D. P. and Moore, D. S. (1980). Chi-square tests of fit for Type II censored data. Annals of Statistics 8, 625-644. [174]

Miller, R. G. (1974). The jackknife - a review. Biometrika 61, 1-15. [182]

_____ (1975). Jackknifing censored data. Technical Report No. 14 (RO1 GM21215), Division of Biostatistics, Stanford University, Stanford, California. [182]

_____ (1976). Least squares regression with censored data. Biometrika 63, 449-464. [150, 163]

Moeschberger, M. L. and David, H. A. (1971). Life tests under competing causes of failure and the theory of competing risks. Biometrics 27, 909-923. [179]

Morton, R. (1978). Regression analysis of life tables and related nonparametric tests. Biometrika 65, 329-333. [146]

Muñoz, A. (1980). Nonparametric estimation from censored bivariate observations. Technical Report No. 60 (RO1 GM21215), Division of Biostatistics, Stanford University, Stanford, California. [177]

_____ (1980). Consistency of the self-consistent estimator of the distribution function from censored observations. Technical Report No. 61 (RO1 GM21215), Division of Biostatistics, Stanford University, Stanford, California. [177]

Nelson, W. (1969). Hazard plotting for incomplete failure data. Journal of Quality Technology 1, 27-52. [67, 165]

_____ (1972). Theory and applications of hazard plotting for censored failure data. Technometrics 14, 945-966. [67, 165]

Oakes, D. (1977). The asymptotic information in censored survival data. Biometrika 64, 441-448. [132]

Peterson, A. V., Jr. (1975). Nonparametric estimation in the competing risks problem. Technical Report No. 13 (R01 GM21215), Division of Biostatistics, Stanford University, Stanford, California. [179]

_____ (1976). Bounds for a joint distribution function with fixed sub-distribution functions: Application to competing risks. Proceedings of the National Academy of Sciences 73, 11-13. [179]

_____ (1977). Expressing the Kaplan-Meier estimator as a function of empirical subsurvival functions. Journal of the American Statistical Association 72, 854-858. [63, 67]

Peto, R. (1972). Discussion on Professor Cox's paper. Journal of the Royal Statistical Society, Series B 34, 205-207. [139]

_____ and Peto, J. (1972). Asymptotically efficient rank invariant test procedures. Journal of the Royal Statistical Society, Series A 135, 185-198. [146]

_____ and Pike, M. C. (1973). Conservatism of the approximation $\Sigma(O-E)^2/E$ in the logrank test for survival data or tumor incidence data. Biometrics 29, 579-584. [117]

_____, Pike, M. C., Armitage, P., Breslow, N. E.,
Cox, D. R., Howard, S. V., Mantel, N., McPherson,
K., Peto, J. and Smith, P. G. (1976). Design
and analysis of randomized clinical trials re-
quiring prolonged observation of each patient.
I. Introduction and design. British Journal of
Cancer 34, 585-612. [117]

_____, Pike, M. C., Armitage, P., Breslow, N. E.,
Cox, D. R., Howard, S. V., Mantel, N., McPherson,
K., Peto, J. and Smith, P. G. (1977). Design
and analysis of randomized clinical trials re-
quiring prolonged observation of each patient.
II. Analysis and examples. British Journal of
Cancer 35, 1-39. [117]

Pettit, A. N. (1976). Cramér-von Mises statistics
for testing normality with censored samples.
Biometrika 63, 475-481. [173]

_____ (1977). Tests for the exponential distribution
with censored data using Cramér-von Mises statistics. Biometrika 64, 629-632. [173]

_____ and Stephens, M. A. (1976). Modified Cramér-
von Mises statistics for censored data. Biometrika 63, 291-298. [173]

Phadia, E. G. (1980). A note on empirical Bayes estimation of a distribution function based on
censored data. Annals of Statistics 8, 226-229.
[80]

Prentice, R. L. (1978). Linear rank tests with right
censored data. Biometrika 65, 167-179. [146]

_____ and Gloeckler, L. A. (1978). Regression analysis of grouped survival data with application
to breast cancer data. Biometrics 34, 57-67.
[139]

_____ and Kalbfleisch, J. D. (1979). Hazard rate models with covariates. Biometrics 35, 25-39. [127, 143]

_____, Kalbfleisch, J. D., Peterson, A. V., Jr., Flournoy, N., Farewell, V. T. and Breslow, N. E. (1978). The analysis of failure times in the presence of competing risks. Biometrics 34, 541-554. [179]

Rai, K., Susarla, V. and Van Ryzin, J. (1980). Shrinkage estimation in nonparametric Bayesian survival analysis: A simulation study. Communications in Statistics, Simulation and Computation B9, 271-298. [79]

Rao, C. R. (1965). Linear Statistical Inference. Wiley, New York. [20, 21]

Reid, N. M. (1981). Influence functions for censored data. Annals of Statistics 9, 78-92. [73, 74, 76, 182]

_____ and Iyengar, S. (1979). Estimating the variance of the median. Unpublished notes. [76]

Sander, J. M. (1975). The weak convergence of quantiles of the product-limit estimator. Technical Report No. 5 (R01 GM21215), Division of Biostatistics, Stanford University, Stanford, California. [76]

_____ (1975). Asymptotic normality of linear combinations of functions of order statistics with censored data. Technical Report No. 8 (R01 GM21215), Division of Biostatistics, Stanford University, Stanford, California. [72, 73]

Schmee, J. and Hahn, G. J. (1979). A simple method for regression analysis with censored data. Technometrics 21, 417-432. [154]

Schoenfeld, D. (1980). Chi-squared goodness-of-fit tests for the proportional hazards regression model. Biometrika 67, 145-153. [175]

Susarla, V. and Van Ryzin, J. (1976). Nonparametric Bayesian estimation of survival curves from incomplete observations. Journal of the American Statistical Association 71, 897-902. [79]

_____ and _____ (1978a). Empirical Bayes estimation of a distribution (survival) function from right censored observations. Annals of Statistics 6, 740-754. [80]

_____ and _____ (1978b). Large sample theory for a Bayesian nonparametric survival curve estimator based on censored samples. Annals of Statistics 6, 755-768. [79]

_____ and _____ (1980). Large sample theory for an estimator of the mean survival time from censored samples. Annals of Statistics 8, 1001-1016. [72]

Tarone, R. E. (1975). Tests for trend in life table analysis. Biometrika 62, 679-682. [118]

_____ and Ware, J. (1977). On distribution-free tests for equality of survival distributions. Biometrika 64, 156-160. [105, 116]

Thomas, D. R. and Grunkemeier, G. L. (1975). Confidence interval estimation of survival probabilities for censored data. Journal of the American Statistical Association 70, 865-871. [52]

Tsiatis, A. (1975). A nonidentifiability aspect of the problem of competing risks. Proceedings of the National Academy of Sciences 72, 20-22. [179]

_____ (1978). A heuristic estimate of the asymptotic
variance of the survival probability in Cox's
regression model. Technical Report No. 524,
Department of Statistics, University of Wisconsin, Madison, Wisconsin. [136]

_____ (1981). A large sample study of Cox's regression model. Annals of Statistics 9, 93-108.
[133, 136]

Turnbull, B. W. (1974). Nonparametric estimation of
a survivorship function with doubly censored
data. Journal of the American Statistical Association 69, 169-173. [7, 57]

_____ (1976). The empirical distribution function
with arbitrarily grouped, censored and truncated
data. Journal of the Royal Statistical Society,
Series B 38, 290-295. [57]

_____ , Brown, B. W., Jr. and Hu, M. (1974). Survivorship analysis of heart transplant data.
Journal of the American Statistical Association
69, 74-80. [141]

_____ and Weiss, L. (1978). A likelihood ratio statistic for testing goodness of fit with randomly
censored data. Biometrics 34, 367-375. [174]

Wilk, M. B. and Gnanadesikan, R. (1968). Probability
plotting methods for the analysis of data.
Biometrika 55, 1-17. [165]

_____ , Gnanadesikan, R. and Huyett, M. J. (1962).
Probability plots for the gamma distribution.
Technometrics 4, 1-20. [166]

Williams, J. S. and Lagakos, S. W. (1977). Models
for censored survival analysis: Constant-sum
and variable-sum models. Biometrika 64, 215-
224. [179]

Zacks, S. and Even, M. (1966). The efficiencies in
    small samples of the maximum likelihood and
    best unbiased estimators of reliability func-
    tions. Journal of the American Statistical
    Association 61, 1033-1051. [35]

Zippin, C. and Armitage, P. (1966). Use of concomi-
    tant variables and incomplete survival informa-
    tion in the estimation of an exponential survi-
    val parameter. Biometrics 22, 665-672. [36]

_____ and Lamborn, K. (1969). Concomitant variables
    and censored survival data in estimation of an
    exponential survival parameter, Part II. Tech-
    nical Report No. 20 (T01 GM00025), Department
    of Statistics, Stanford University, Stanford,
    California. [36]

INDEX

S